THE ENGINEER AND HIS PROFESSION

2ND EDITION

HOLT, RINEHART AND WINSTON

New York Chicago San Francisco Atlanta
Dallas Montreal Toronto London Sydney

John Dustin Kemper
University of California, Davis

THE ENGINEER AND HIS PROFESSION

2nd Edition

Library of Congress Catalog Number: 74–15607
ISBN: 0–03–088348–2
Printed in the United States of America
9 8 7 6 5 4 3 2 1 090 8 7 6 5

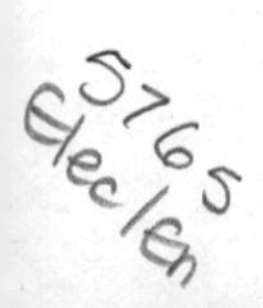

To my wife, Barbara

PREFACE

This book is intended primarily for students who are preparing themselves to become engineers and for those who have recently entered the profession. Nevertheless, all engineers, whether young or old, should find something of interest herein. Some of the material will be regarded as controversial because it reflects the nature of our time; however, I have tried to be as factual as I can, and have endeavored to present both sides of controversial issues.

In the 1967 edition of this book, I wrote that the engineer was devoted to the ideal of technical advancement, and to a large degree this ideal had also been adopted by the general public. I then said:

> Technical advancement may be the guiding devotion of the engineering profession, but the members of this group should be fully aware that *human welfare* is the ultimate justification for technology's existence. Regardless of whatever views the reader may hold concerning the workings of society, history shows that no society will long tolerate any group within itself whose aims are regarded as antagonistic to those of the majority.

These words seem to me to be even more vital today than when they were first written. In the intervening time, we have experienced events that have wrenched the nation and the world. We have landed men on the moon—not once, but many times. The Viet Nam war has escalated and deescalated. We have experienced a bewildering succession of one crisis after another: the identity crisis, the environmental crisis, the credibility crisis, and the energy crisis.

Withal, regrettably, the public does not have the confidence in technology that it once did. I think we can no longer say that the general public gives its unqualified devotion to technical advancement. This much is probably all right, for we are surely all in trouble if we blindly give our devotion to technical advancement. The unhappy part is that much of the nation, and perhaps many engineers, too, have lost a great deal of confidence not only in technology, but in the effectiveness of our social institutions in serving our needs. The serene (and unwarranted) confidence that engineers and scientists can do anything, given enough money, has yielded to a fear expressed by some (also unwarranted) that engineers lack any real interest in social improvement, but are the captives of a greedy something called the Establishment.

There is little doubt in my mind that the three great issues facing the engineering profession—indeed, facing all of mankind—are 1) the environment, 2) ethical behavior, and 3) resources, especially energy resources. Hence, in writing a new edition, it was clear that these matters would have to be dealt with in a major way. New sections have been added on the environment, energy, public responsibility, and manpower. Most of the material of the first edition has been retained, with up-dating of some items and condensation of others.

Regarding the environmental and ethical issues, I have sought enlightenment from many sources. For twenty years, I have been a member of the Sierra Club, and have absorbed the environmental views of that organization. I have in a small way helped in the fight for saving wilderness, for national parks, for open space, and for coastline preservation. Beyond this, I have tried doggedly to keep up with the astonishing avalanche of environmental literature. On ethical issues, I have sought enlightenment from the philosophers, and have tried to understand the thoughts of the more virulent critics of American society—Marcuse, Roszak, Reich, Nader, and others.

I have also tried to gather views from other kinds of sources—from government publications, from the establishment press, and from that tiny percentage of environmental books which take a more positive view of environmental problems than does the majority.

In addition, to the degree possible for any one individual, I have tried to check what I have read against my own perceptions. When, for example, I read in the literature that American society feels betrayed and alienated, I observe that I certainly know *some* who are alienated. But the overwhelming majority of those I meet seem to feel pretty much as I do—that life holds promise, that progress is being made, and that society, though possessing some of the traits of a dinosaur, is groping toward a more humane state.

When I read that America's forests are being destroyed, my actual perceptions in the field are that, quite to the contrary, vast tracts of forests are being managed carefully by the Forest Service, and that huge lumbering companies are staking their futures on investments in nurseries, seed orchards, and replanting programs that will not yield returns until the next generation. When I read that vast tracts of land are being permanently destroyed by strip mining, I find instead that many of those lands have already been reclaimed, largely by natural processes.

I have tried personally to visit some of those places where environmental degradation is supposed to be the worst. I did not have to visit Los Angeles to check on its air pollution. After living there for fifteen years, I know there is an air pollution problem in that city. I have visited the hills of West Virginia and Kentucky, have seen the damage wrought by strip mining, but have also seen that reclamation of strip-mined lands *can* work—that it is not the empty gesture some have alleged. I have hiked and camped in many of America's wild places, and have observed that things there are neither as bad as has been claimed, nor as good as we would like. I have seen the clear-cutting of the redwoods, but have also seen the return to virgin conditions of regions that were clear-cut a century ago.

To say, as some have, that the human race has only a few more years left before it self-destructs is the most antihumanity thing I have ever heard. The biggest lesson in history is that humanity possesses enormous reserves of adaptability and creativity to help it solve its problems. I don't know how long the human race is destined to survive on this planet, but I'll wager it will be a long time. Furthermore, I strongly suspect that the crises of today will be eclipsed by the crises of tomorrow, just as the agonies of the Great Depression and of World War II are largely forgotten under the pressure of today's anxieties.

I suppose it is only fair to admit, right at the outset, that I am a technological optimist. I believe that technology, intelligently applied, is our only effective route to environmental improvement. I think we have been far too short-sighted in assessing the true state of affairs as regards our environment. We have forgotten that more of us are better-fed, better-housed, and better-informed today than ever before in history, and we tend to overlook the fact that these are extremely important attributes of the total human environment. We don't want to lose our grasp on these precious achievements while pursuing the goals of cleaner air, cleaner water, and natural esthetics. We want *all* these things, and it is only through the intelligent use of technology that we can succeed, in my opinion.

Engineering is an enormous and a highly diversified field. It may be that not every engineer will find his or her particular kind of engineering included in this book. The aspects of engineering that are treated in most detail are those of research, design, and development, together with related environmental, ethical, resource, and management issues. In my opinion, it is these matters that go straight to the foundation of the engineering profession.

Because management is likely to be one of the areas of greatest mystery and misinformation for young engineers, this book dwells at some length on the subject. Many engineering students want to go into management, and indeed, a large percentage of them do wind up in some kind of management activity. However, many other young people are uncertain, and even confused, regarding careers in management. It is hoped that this book will help to resolve some of their uncertainties.

I am deeply indebted to my colleagues, especially those listed below, who have had a part in shaping my ideas concerning the engineering profession and its role in human society. Many of them have taken their valuable time to read portions of the manuscript and to make suggestions for improvement. They may not all be in complete agreement with everything I have said; but if the book is better than it might otherwise have been, the credit is theirs: James W. Baughn, Charles W. Beadle, Richard L. Bell, Harry Brandt, John W. Brewer, Donald G. Crosby, James A. Cheney, John R. Goss, Armand G. Guibert, Jerald M. Henderson, S. Milton Henderson, Leonard R. Herrmann, Harry R. Kattelmann, Jack W. La Patra, Richard B. LeVino, Coby Lorenzen, Herschel H. Loomis, Jr., Clarence W. Martin, William L. Martin, Allan A. McKillop, Alan T. McDonald, Earle W. Owen, Karl M. Romstad, G. Scott Rutherford, Wilson K. Talley, Arthur J. Winter, and An Tzu Yang. My thanks also go to Eldora Synhorst and Wanda Winton, who typed the manuscript.

I would like to express a special debt of gratitude to the following men who have had a deep impact upon the development of my ideas concerning engineering and engineering education. All of them have been or currently are deans or associate deans of engineering at the different campuses of the University of California: Roy Bainer, Don O. Brush, the late Llewellyn M. K. Boelter, Clyne F. Garland, Warren H. Giedt, Ray B. Krone, Morrough P. O'Brien, John B. Powers, and John R. Whinnery.

Davis, Calif.
January 1975

J. D. K.

CONTENTS

THE ENGINEER
AND HIS PROFESSION
2ND
EDITION

ONE

The environment

Engineers, whether knowingly or unknowingly, are to a very great degree the architects of our environment. They design the structures within which we live, create the industrial plants upon which we depend, and invent the gadgets which form a part of our everyday existence. They have also created some unwanted components of our environment. The structures have sometimes helped cities to be shaped in undesirable ways, as when freeways have permitted commuters to escape to the suburbs and thus abandon city cores to decay. The industrial plants have polluted the atmosphere and contaminated the waterways. The gadgets have created whole revolutions in societal forms, such as instant news of worldwide events brought by radio and television or a mobile and perhaps rootless society produced by the automobile.

If the engineer is to play such an influential role in shaping our environment, he should play that role consciously rather than unconsciously. This is not to say that engineers of the future must never make any mistakes in assessing the environmental consequences of their acts. Such a demand cannot reasonably be made of individuals who are, after all, human. But it does suggest that far fewer mistakes will be made by engineers who are fully aware that their daily decisions have important consequences. Engineers who are constantly looking for environmental improvements are much more likely to find them than those who do not look.

In the 1960s, as the public became aware of its pollution problems, it seemed that this issue would unify society. There was general alarm over air pollution, water pollution, pesticides, junkyards, billboards, and damage to the wilderness; new environmental organizations were formed, and old ones gained strength; new laws were enacted; books on the environment were written and published.

The illusion of unity did not last. Radicals seized upon environmental pollution as a sign of establishment breakdown. Minority groups charged that their legitimate interests were being forgotten in the rush. They pointed out that the opportunity to be employed in a gainful occupation is the most basic environmental ingredient of all. Of what use, they said, is it to preserve wilderness or to clean up Lake Erie, if millions of Americans must live in poverty without jobs. Jobs should be top priority—not the environment. Finally, to complete the picture, members of the corporate establishment declared that the Gross National Product must be kept rising at all costs—that the Clean Air Act should be put on the shelf and the Environmental Protection Agency placed in neutral.

The loss of apparent unity was regrettable but probably inevitable. Environmental concerns are linked to some of the most basic aspects of human affairs. An effort to improve air pollution, for example, was found to be tied to economic productivity. If productivity goes down, unemployment and poverty will rise.

Voices cried out against technology itself as the villain. "We want no more technology," said some. "Technology will only make matters worse." For a time, such a view seemed to be gaining in strength, especially on university campuses. But for the most part the mood passed and gave way to a realization that technology, applied with care, is a vital ingredient in our battle for environmental improvement. Even Barry Commoner, the prominent antitechnologist, proved not to be so antitechnology after all. In his best seller, *The Closing Circle*, Commoner says: "If we are to survive economically as well as biologically, industry, agriculture, and transportation will have to meet the inescapable demands of the ecosystem. This will require the development of major new technologies. . . ."[1] He even suggests specific technological solutions for certain problems when he proposes that we should construct new urban collection systems for sewage and build pipelines to carry it to the countryside for incorporation into the soil.

Engineers should have a basic understanding of environmental matters, since they are simultaneously being blamed for the existence of pollution and also are expected to do something about it. Unfortunately, most of the popular environmental literature is of little help in shaping an informed opinion on the state of environmental affairs. It consists mostly of passionate calls to political action, frequently based upon speculative premises. For example, in recent years two contradictory theories have emerged, each predicting diametrically opposite results. In one theory, the combustion of fossil fuels is predicted to increase the CO_2 concentration in the atmosphere to such a degree that the heat balance of the earth would be upset. More heat would be retained by the earth than formerly because of the "greenhouse effect" of the increased CO_2, melting the polar ice caps and flooding the coastal plains of the world. In the other theory, the combustion of fossil fuels would increase the concentration of particulates in the atmosphere, reflecting more of the sun's rays back into space. The earth would grow

---

[1] B. Commoner, *The Closing Circle* (New York: Bantam Books, 1972), p. 282.

colder, bringing on a new ice age.[2] Both theories are actually only speculations. Proponents of the ice-age theory cited recent cooling trends in the earth's climate over the past few decades to support their views. But this is no proof at all, because the earth's climate has undergone several marked shifts within historical times. There have been extended dry periods and extended wet periods. Glaciers in Europe have retreated, advanced, and then retreated only to advance again. In the early seventeenth century, glaciers in the western Alps advanced to the point where they overran villages that had been inhabited for centuries.[3] To appeal to short-term climatic changes as supporting evidence for an environmental theory is thus untenable, although some scientists are now predicting it is about time for the earth to enter another ice age on a normal cyclic basis. However, the process of doing so would probably take several thousand years.[4] At a 1972 conference at Brown University, a group of scientists interested in interglacial periods concluded:

> . . . there is no qualitative difference between the climatic fluctuations in the 20th century and the climatic oscillations that occurred before the industrial era. The present climatic trends appear to have entirely natural causes, and no firm evidence supports the opposite veiw.[5]

Another example worth mentioning is the cry of alarm which arose from those who feared we would seriously deplete the earth's supply of oxygen because of the ever-growing practice of burning fossil fuels. But a group of scientists and engineers who studied this and related matters concluded that it was a "nonproblem." They found no changes in atmospheric oxygen since the time when regular measurements began in 1910. Further, they calculated that if all the known fossil fuels in the world were consumed, the consumption of oxygen caused thereby might reduce the oxygen concentration in the atmosphere from 20.9 percent to 20.8 percent.[6]

Is there truly an environmental crisis? The question must be asked, for it is wasteful, and perhaps even harmful, to take crisislike actions if our perceptions of crisis are wrong. Is there, for example, a water pollution crisis? To read the newspapers, one would have no doubt: our rivers are all running sewers of chemicals and filth. Yet we drink

---

[2] L. Rocks and R. P. Runyon, *The Energy Crisis* (New York: Crown Publishers, 1972), pp. 118–119.

[3] C. E. P. Brooks, *Climate Through the Ages* (New York: Dover Publications, 1970)

[4] "A New Ice Age?", *Mosaic*, Spring 1973, p. 23.

[5] *Science*, October 13, 1972, p. 190.

[6] *Man's Impact on the Global Environment: Report of the Study of Critical Environmental Problems* (Cambridge, Mass.: The MIT Press, 1970), pp. 74–75.

purer water today, by and large, than we did 100 years ago. In that prior era we genuinely had a water pollution crisis. For example, in 1849 and 1853, 20,000 people in London were killed by cholera. As Donald Carr says, "In the Western world this was the greatest pollution disaster of history."[7] Typhoid and cholera epidemics were widespread, stemming from water contaminated by sewage. During the great wagon train emigration of 1849, more deaths occurred from cholera than from Indians, wild beasts, and shootings. The greatest danger to the emigrant came as he waited alongside the polluted rivers of Missouri to begin his journey to the west.[8] In comparison with such a lurid past, it is hard to say whether we have a water pollution crisis today or not. We certainly have serious water pollution problems, but it is difficult to document a state of crisis except in a few localized areas.

Pure air

With air pollution, there can be little argument that conditions have reached a state of crisis in some cities. The difference between air and water which makes pollution of one a crisis and of the other not is simply this: we live within an ocean of air; we do not live in the water. We cannot escape our need for air even for a minute. With contaminated water, we have a little more time, during which we can seek alternate sources to sustain us while we solve pollution problems. With air we cannot do so.

The biggest cost of air pollution is to human health. Other aspects, such as loss of visibility and unpleasant smell, are important but are relatively insignificant when compared to health. The major components of air pollution are described below, together with their principal effects.

SULFUR OXIDES.[9] Sulfur dioxide (SO_2), sulfur trioxide (SO_3), and their corresponding acids, sulfurous acid (H_2SO_3) and sulfuric acid (H_2SO_4), are among the oldest and best known atmospheric pollutants. They arise principally from the burning of fossil fuels such as coal and petroleum, although many other kinds of industrial activity also produce sulfur oxides. SO_2 is a gas which reacts with oxygen in the atmosphere to form SO_3, which then combines immediately with water to yield H_2SO_4 in the form of droplets. Within the United States, the highest SO_2 values have been reported in the northeastern region, where

---

[7] D. E. Carr, *Death of the Sweet Waters* (New York: Berkley Publishing, 1971), p. 41.

[8] D. M. Potter, *Trail to California* (New Haven: Yale University Press, 1945), p. 54ff.

[9] *Air Quality Criteria for Sulfur Oxides* (Washington, D.C.: National Air Pollution Control Administration, January 1969).

large quantities of high-sulfur fossil fuels have been burned. For most large U.S. cities, the pollution from SO_2 has improved markedly in recent years, mostly because of shifting from high-sulfur to low-sulfur fuels.

Sulfur oxides are clearly detrimental to plant life, with severe plant damage having been observed miles downwind from certain smelting operations. They produce corrosion in metals, discoloration of fabrics, and deterioration of building materials. A combination of sulfur oxides and particulates seems to be especially damaging to human health, partly because of the action of small particles in conveying H_2SO_4 into the lungs, and partly because of the chemical role of particulates in converting SO_2 to H_2SO_4. Dramatic episodes, involving high mortality, have occurred in which sulfur oxides and particulates have figured prominently.

The world's most disastrous smog incident occurred in London in December, 1952, causing about 4000 excess deaths over those which would normally occur. Elderly persons and those with preexisting pulmonary and cardiac disease were most susceptible. The maximum daily concentration of SO_2 recorded during the 1952 London smog was 1.34 parts per million (ppm). The Environmental Protection Agency (EPA) has recommended that the annual mean SO_2 concentration be no greater than 0.02 ppm; from 1963 to 1970, the annual mean in the United States for 21 key cities declined from about 0.027 ppm to about 0.013 ppm.[10]

Numerous potential methods are known for removing SO_2 from stack gases.[11] All of them involve extra expense. A major problem exists in deciding whether to use a throw-away process or one which attempts to recover the sulfur. In the throw-away case, the need for disposal of a waste product creates a new problem. On the other hand, attempts to recover and sell the sulfur can create major dislocations in the market.

PHOTOCHEMICAL OXIDANTS.[12] Even in the absence of sulfur, combustion of fossil fuels causes serious air pollution in urban atmospheres. Since combustion processes are less than 100 percent efficient, exhaust gases usually contain unburned hydrocarbons (HC), as well as nitrogen oxides (NO_x) and the "normal" combustion products such as CO_2 and H_2O. In the atmosphere, many of these products react chemically to produce new contaminants. These processes, which are not fully understood, are stimulated by sunlight, and the products are thus termed generally *photochemical oxidants*. Two of the principal photo-

---

[10] *Environmental Quality* (Washington, D.C.: The Council on Environmental Quality, August 1972), p. 7.

[11] A. V. Slack, "Removing SO_2 from Stack Gases," *Environmental Science and Technology*, February 1973, pp. 110–119.

[12] *Air Quality Criteria for Photochemical Oxidants* (Washington, D.C.: National Air Pollution Control Administration, March 1970).

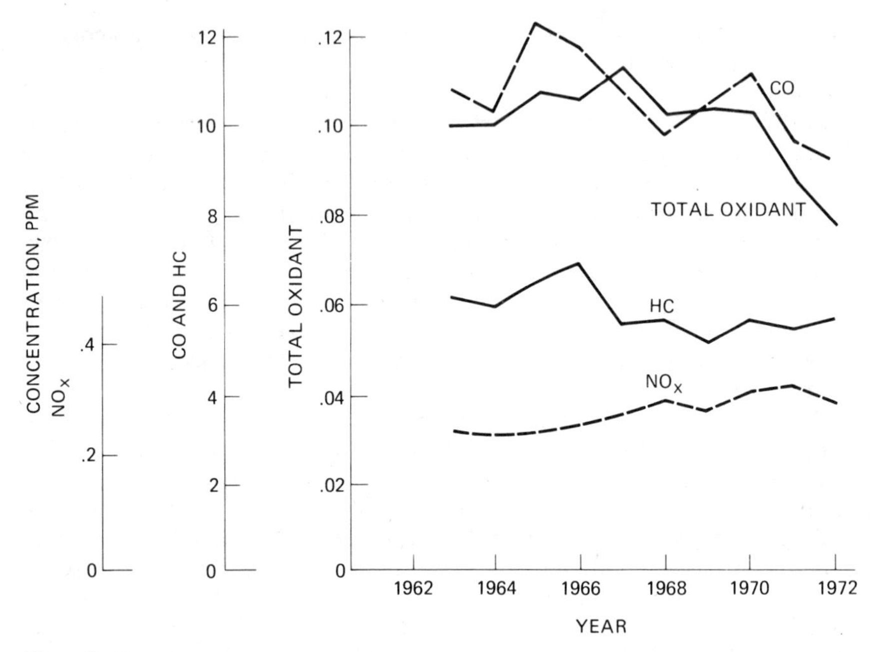

Fig. 1–1 Trend lines of four major air pollutant concentrations in the Los Angeles area, in parts per million (ppm). The pollutants are total oxidant, carbon monoxide (CO), hydrocarbons (HC), and oxides of nitrogen (NO_x). Data are given for yearly averages of maximum average concentration reached during any hour of each day. (Source: *Air Pollution in California, 1972*, Air Resources Board.)

chemical oxidants are ozone and peroxyacetyl nitrate (PAN). (Ozone is also continually being created in the atmosphere by natural processes, but not to a degree great enough to constitute a pollution hazard.) This type of polluted air is often referred to as photochemical smog. The highest concentrations of photochemical oxidants in the United States are found in the Los Angeles area, although all large cities in the world are afflicted with smog. The California Air Resources Board, however, was able to report in 1973 that Los Angeles was showing slow improvement in its smog problem.[13] Concentrations of HC, carbon monoxide (CO), and total oxidants in Los Angeles' atmosphere have been declining since the peak year of 1967, principally because of control of motor vehicle exhausts (see Figure 1-1). Oxides of nitrogen, however, were still increasing.

---

[13] *Air Pollution in California, 1972* (Sacramento: Air Resources Board, State of California, January 1973).

The damaging effects of photochemical oxidants are many. Vegetation has been seriously affected in parts of California, for example, with a serious decline in citrus fruits and other crops reported from the Los Angles area. Materials are adversely affected by high concentrations of ozone, especially rubber.

Insofar as human health is concerned, the data concerning the effects of photochemical smog are mixed and inconclusive. Clearly, eye irritation is a major effect of smog. Also, impairment of performance by high school athletes has been observed when concentrations of photochemical oxidants were 0.3 ppm. (Peak concentrations in the Los Angeles area have gone as high as 0.67 ppm.) Significantly, more auto accidents occur on days of high oxidant concentrations, and surveys show a higher frequency of asthma, coughs, and nose and throat complaints in southern California than in the rest of the state. However, no clear association has been established between photochemical air pollution and increased death rate in the fashion that was possible for sulfur oxides.

HYDROCARBONS.[14] No direct health effects of HC concentrations on urban populations have been demonstrated. Most of the principal effects of hydrocarbons are caused by their photochemical products.

It should be realized that not all HC in the atmosphere results from human activity. For example, large amounts of methane (CH_4) are generated naturally by decomposition of organic matter at the earth's surface. Also, volatile terpenes and isoprene are emitted by some kinds of vegetation. Natural hazes associated with vegetation occur in many areas of the world. An example is the blue haze of the Appalachian Mountain region. No doubt many people, suddenly becoming conscious of such a haze, have assumed it is the consequence of man's activity and may say, "Look at that smog!" However, the concentrations of such natural emissions are too low to be of concern in urban areas.

The principal sources of the troublesome HC components are the various combustion processes, especially those in internal combustion engines. Evaporation of solvents is also an important HC source as is the evaporation of gasoline at filling stations. In addition to the control of HC in motor vehicle exhausts, proposals have been advanced to require that filling stations convert to completely closed systems, so that no gasoline fumes escape into the atmosphere.

CARBON MONOXIDE.[15] This pollutant is deadly when encountered in sufficient concentration. CO is produced by incomplete combustion of

---

[14] *Air Quality Criteria for Hydrocarbons* (Washington, D.C.: National Air Pollution Control Administration, March 1970).

[15] *Air Quality Criteria for Carbon Monoxide* (Washington, D.C.: National Air Pollution Control Administration, March 1970).

carbonaceous materials, principally in internal combustion engines. Generally speaking, CO production in engine exhausts is a function of the air/fuel ratio. Rich mixtures produce more CO (and also more unburned hydrocarbons) than do lean mixtures. Unfortunately, it is not feasible to solve our pollution problems simply by using lean mixtures, because as the CO and HC decline with progressively leaner mixtures, the nitrogen oxides (NO_x) increase.

Total emissions of CO probably exceed those of all other pollutants combined according to the National Air Pollution Control Administration (NAPCA). Most of these emissions come from technological sources, although CO is also created to some extent by natural processes. Overall atmospheric levels of CO do not seem to be rising over the years, even though such a large amount is being produced annually that concentration should be rising steadily. Hence, there must be some important natural removal process at work, such as oxidation of CO to CO_2 in the upper atmosphere, or absorption by plants and microorganisms. In fact, it has been suggested by Dr. Ray B. Krone, an environmental engineer at the University of California, Davis, that a contributing reason for air pollution in urban areas is that so much of the surface is covered by pavement that interaction of the atmosphere with the soil and its microorganisms is prevented, which interferes with natural purifying processes.

Deaths from CO poisoning in poorly ventilated rooms, cars, and garages are well understood. What are not so well understood are the effects of exposure to relatively low concentrations. Short-term exposures to moderate concentrations are known to produce deterioration in vision and muscular action. The effects of long-term exposure to low concentrations are not known, but concerns have been expressed by NAPCA that permanent impairment of the cardiovascular and nervous systems could be caused by continual exposure to CO. Attempts to discover a relation between physical damage and exposure to CO for groups such as traffic policemen, parking garage attendants, and emloyees of New York's Holland Tunnel have all been inconclusive. Similarly, attempts to relate CO concentrations to increased likelihood of motor vehicle accidents have been clouded by uncertainties. One complicating factor is that a regular smoker may have as much CO in his blood from smoking cigarettes as if he had been exposed for hours to such a poorly ventilated environment as a parking garage or automobile tunnel.

NITROGEN OXIDES.[16] There are several known oxides of nitrogen, but the important ones from the standpoint of air pollution are nitric oxide (NO) and nitrogen dioxide (NO_2). Common practice is to refer

---

[16] *The Oxides of Nitrogen in Air Pollution* (Berkeley: State of California, Department of Public Health, January 1966).

to nitrogen oxides collectively as NO_x. NO_x is a product of almost any combustion process which uses air, since nitrogen is the chief component of air. To a great degree, the formation of NO_x is the result of high combustion temperatures. The principal sources are motor vehicles which account for as much as 50 percent to 60 percent of NO_x in the atmosphere in industrialized urban areas.

The nitrogen oxides participate actively in photochemical reactions with hydrocarbons, thus helping to produce photochemical smog. NO_2 plays a double role in air pollution, one as a component in the formation of photochemical smog and the other as a toxicant in its own right. NO is much less toxic than NO_2, but NO is readily converted into NO_2 in the atmosphere by reaction with oxygen.

Laboratory tests on animals and plants have shown that NO_2 has pronounced toxic effects if present in sufficient concentrations. In general, the evidence of toxic effects at concentrations encountered in urban atmospheres is inconclusive. There is concern, however, that the concurrent presence of particulates along with various pollutants may have the effect of concentrating damaging chemical action in the lungs. In any event, even if the direct toxicity of NO_2 is discounted, its role in the formation of other photochemical smog products causes it to be numbered among the important pollutants.

Nitrogen oxides were brought within the realm of air pollution control much later than other pollutants because of uncertainty about their damaging properties. Emission of NO_x from cars produced in the late 1960s actually increased over that of earlier models as a consequence of the methods used to control HC and CO emissions.[17]

As an example of the uncertainties which occur in pollution control, we should consider the embarrassment of the Environmental Protection Agency in 1973, when it announced that the measuring techniques it had been using for NO_x were faulty. As a result, atmospheric NO_x levels had been overestimated, and the EPA said the problem did not appear to be as widespread or as serious as previously believed. Nevertheless, the EPA said standards for NO_x were still necessary to protect public health, although the control measures needed might not be as stringent as previously believed.

PARTICULATES.[18] Particulate matter in the atmosphere consists of a multitude of substances. Collectively, these particulates are referred to as *aerosols*, whether in the form of solid particles or liquid droplets. Typical components are such things as dust, carbon particles, salt crystals, metal grindings, cement powder, and sulfuric acid mist. All sorts of industrial processes contribute to the total particulate matter in the

---

[17] *Air Pollution in California, 1972, op. cit.*, p. 9.

[18] *Air Quality Criteria for Particulate Matter* (Washington, D.C.: National Air Pollution Control Administration, January 1969).

atmosphere, especially those involving combustion. Agricultural burning is also a major source of particulates, and much controversy is involved on this topic. Farmers point out, with some justification, that if they are required to dispose of their wastes by more costly means than open burning, a rise in food prices is inevitable.

Nature also accounts for much particulate matter in the air. In the 1930s, enormous quantities of dust were carried into the air from the midwestern "Dust Bowl" states; some of it was carried 2000 miles to New York City—enough to cause twilight conditions. In the early 1900s, mountaineers who climbed the peaks of the Pacific Northwest repeatedly found they were unable to enjoy the magnificent views they had anticipated, because smoky palls from forest fires covered the region.[19] In 1883, volcanic ash from the explosion of the Krakatoa volcano near Java was spread throughout the atmosphere of the earth, persisting for years.[20] Volcanic eruptions are frequent events. From 1968 through 1971, 71 eruptions at various places in the world were reported.[21]

Evidence of the direct toxic effects of particulates on humans is meager, with the exception of combination with H_2SO_4. Great concern has been expressed over the amounts of toxic materials such as lead, beryllium, and asbestos being discharged into the atmosphere. No evidence is available that they presently constitute a serious risk to the population at large, but they could become serious if their concentrations increase. Particles may play a special role, however, for they may adsorb other toxic substances on their surfaces and then transport these concentrated toxicants into the lungs. Lead, especially, is a problem near freeways. Over 500 millions pounds of lead are burned per year in motor fuel and discharged to the atmosphere.[22]

One of the most dramatic modern cases of cleaning up a polluted city is that of Pittsburgh, Pennsylvania. Prior to the 1940s, Pittsburgh residents often could not see across the street because of the pollution in the air. By the 1960s, Pittsburgh had become one of America's cleaner cities. The improvement did not come easily. Political battles were fought and then fought again over the issue of whether the cost of pollution control would drive industry from the city and turn it into a ghost town. By the mid-1960s hundreds of millions of dollars had been spent on pollution control, but other hundreds of millions had been saved on cleaning bills alone, without attempting to include a dollar

---

[19] C. E. Rusk, *Tales of a Western Mountaineer* (Boston: Houghton Mifflin, 1924).

[20] L. Bertin, *The New Larousse Encyclopedia of the Earth* (New York: Crown, 1972), p. 47.

[21] J. Cornell and J. Surowiecki (Eds.), *The Pulse of the Planet, A State of the Earth Report from the Smithsonian Institution Center for Short-Lived Phenomena.* (New York: Crown, 1972).

[22] *Project Clean Air*, Vol. 2 (Berkeley: University of California, September 1, 1970), pp. 2–23.

value for the improvement in human health. Visibility had improved 77 percent over 1945, and far from shutting down, the city's steel mills were producing at record levels.[23]

Many people believe that little has been done to control smog, and that any progress has been largely overrun by population increases and by increased technology. In California, where photochemical smog is an acute problem, the state's Air Resources Board states flatly that this is not true.[24] It points, for example, to improvements in the Los Angeles situation. The Board also offers its prediction that even greater improvement can be expected by the 1980s, based upon the growing use of control technologies and the retirement from use of older motor vehicles. If the Air Resources Board's predictions are correct, the motor vehicle by then will no longer be the prime culprit in Los Angeles' smog, and in the Board's language, ". . . greater emphasis will have to be placed on emissions from nonvehicular sources."[25]

Pure water

In the case of water, we must still be concerned with direct health effects upon humans, but we also must be concerned with the health of organisms which live within the water. This is partly because we use some of these organisms for food and partly because of various subtle but important indirect effects these organisms may have upon human welfare.

We also must be concerned with the esthetic and recreational qualities of water. Surveys have shown that Americans mostly require water in their outdoor recreation, ". . . to sit by, to swim and to fish in, to ski across, to dive under, and to run their boats over."[26] The recreational use of water has assumed a high priority as the nation has undergone a transition from an agrarian to an industrial and urban society. This transition has brought with it increasing amounts of leisure time and a serious concern for the creative uses of this leisure.

Agriculture is one of our biggest polluters because of the volumes of pesticides and nutrients from fertilizers which drain off farm lands and into surface waters. Industry produces unwanted nutrients, too, besides discharging many highly toxic substances into waterways. Agriculture and industry are also huge *users* of water, and have stringent quality re-

---

[23] T. O. Thackrey, "Pittsburgh: How One City Did It," in *Controlling Pollution—The Economics of a Cleaner America,* M. I. Goodman (Ed.) (Englewood Cliffs, N.J.: Prentice-Hall, 1967), p. 139.

[24] *Air Pollution Control in California, 1971* (Sacramento: Air Resources Board, State of California, January 1972).

[25] *Ibid.*, p. 34.

[26] *Outdoor Recreation for America* (Washington, D.C.: Outdoor Recreation Resources Review Commission, 1962), p. 4.

quirements on the water they use regarding salinity, acidity, chemical content, dissolved minerals, and suspended solids.

Broadly speaking, the aspects which concern us most in water pollution are microorganisms, oxygen content, excess nutrients, pesticides, and toxic materials such as the heavy metals.

From the standpoint of domestic usage, the most prominent concern in water pollution is the capacity for transmitting disease through microorganisms. As mentioned previously, some of the greatest pollution disasters in history have been cholera and typhoid epidemics caused by contaminated water. Even in modern times, typhoid and cholera account for untold numbers of deaths in some countries of the world, and in the United States, infectious hepatitis from contaminated water remains a problem. Raw sewage from 20 million people was still being discharged into U.S. streams in the mid-1960s.[27] And, unfortunately, a community with modern sewage treatment facilities may find that its treatment plant frequently does not function properly because of overloading, or because of a sudden discharge of industrial waste which upsets the plant's balance. The greatest single category of federal funding for environmental improvement is for construction of waste treatment plants, accounting for $1.65 billion in 1972.[28] In spite of the astonishing amount of sewage pollution in the United States, the country still enjoys remarkably clean water, achieved through filtration and chlorination of municipal water supplies, and water-borne disease is a relative rarity.

One of the important aspects of water quality, other than microorganisms, is its dissolved oxygen content, for without oxygen most forms of aquatic life cannot survive. The discharge of nutrients into waterways has an important impact on dissolved oxygen, because the nutrients feed dense blooms of algae. The algae blooms in themselves may be obnoxious, and some kinds of algae are toxic to humans.[29] But algae also have the additional effect of consuming the oxygen in the water during decomposition after death. Various kinds of bacteria are nourished by nutrients coming from raw domestic sewage and industrial effluents. These create tremendous oxygen demands upon the water in which they live and are also noxious to many creatures, including man.[30]

Much of the trouble in the highly publicized case of Lake Erie has been caused by the depletion of oxygen in the lake, with resultant adverse impacts on fish life. Lake Erie fish populations have also been seriously reduced by overfishing and by forms of pollution other than

---

[27] *Death of the Sweet Waters, op. cit.*, p. 42.

[28] *Environmental Quality, op. cit.*, p. 119.

[29] *Water Quality Criteria* (Washington, D.C.: Federal Water Pollution Control Administration, April 1, 1968), pp. 51–56.

[30] *Ibid.*, p. 97.

oxygen depletion. In fact, it has become popular in some circles to say that "Lake Erie is dead." It is not dead, of course, and is actually slowly recovering as a consequence of pollution control programs. Full recovery cannot be expected, however. In fact, "full recovery" may not even be desirable. As two scientific investigators of the lake's ecology have stated, ". . . if, for example, [full recovery] meant the return of mosquitoes in their original numbers, with the attendant epidemics of malaria, few of us would choose it."[31]

The measures needed to prevent the discharge of nutrients into lakes may require strong action. California's Lake Tahoe, for example, is a mountain jewel with water so clear that objects can be seen at great depths. However, massive recreational developments in the region have caused great amounts of nutrients to be carried into the lake from sewage systems and septic tanks. As a result, regular algae blooms have begun to appear in the formerly clear waters. To correct this problem, a new sewage system has been constructed which transports the treated water into adjacent drainage basins for discharge. This additional water is actually welcome in these adjacent basins for irrigation purposes, because the lands there are largely water-deficient.[32]

Pulp and paper mills are big polluters. Much of the waste from such mills is biodegradable by bacterial action. Unfortunately, the degradation process imposes a large biochemical oxygen demand (BOD) upon the water, which depletes its oxygen and suffocates aquatic life. One method of reducing the BOD of pulp-mill wastes is to induce biological degradation before discharge into surface waters.[33]

Two of the principal nutrients involved in water pollution are nitrogen and phosphorus. Sources of both are to be found in domestic sewage, industrial discharges, animal and plant processing wastes, and agricultural runoff. The Council on Environmental Quality reported in 1972 that the problem of nitrogen and phosphorus in the United States was worsening dramatically, probably because of increased use of fertilizer.[34] Nitrogen can also enter surface waters directly from the atmosphere and from animal manure. The latter source has become progressively more important as the tendency to concentrate animals in mass production facilities has grown.

A noteworthy source of phosphorus in recent years is detergents. These soap substitutes attracted attention in the 1960s because they created mounds of foam in sewage outfalls and sometimes even made their way into domestic water supplies. (One may understandably be

[31] H. A. Regier and W. L. Hartman, "Lake Erie's Fish Community: 150 Years of Cultural Stresses," *Science*, June 22, 1973, p. 1248.

[32] R. L. Culp and H. E. Moyer, "Wastewater Reclamation and Export at South Tahoe," *Civil Engineering—ASCE*, June 1969, p. 38ff.

[33] *Water Quality Criteria, op. cit.*, p. 90.

[34] *Environmental Quality, op. cit.*, p. 13.

disturbed if a foamy head of suds appears on a glass of water freshly drawn from the tap.) The detergent industry in 1965 switched to a new type of detergent with improved biodegradable qualities and less tendency to foam after passing through treatment plants. The new detergents still contained phosphorus, however, and most of the phosphorus continued to pass into surface waterways, contributing to biological growths and imposing a large BOD on the water.[35] Late in the 1960s, under government pressure, the detergent industry attempted to do away with phosphorus altogether and switch to a phosphate substitute known as sodium nitrilotriacetate (NTA). This move dissolved into controversy when it began to appear that NTA was a hazard to human health, whereas phosphates were not. It was also pointed out that detergents represented only one of the many sources of phosphorus entering surface waters, and that the best approach might be to apply newly developing technology to remove phosphorus in sewage treatment plants, before discharge into waterways.[36]

An obvious requirement of pure water is that it be free of toxic materials such as the heavy metals. The number of potential poisons is very large, and their effects on animals are complex. The matter is further complicated by the fact that some materials, such as copper, are necessary to life in trace amounts but are poisonous at heavier concentrations. Dr. Donald G. Crosby, an environmental toxicologist at the University of California, Davis, has pointed out that we are being subjected to all sorts of poisons at all times, most of them coming from natural sources, but our physiological systems have evolved over millions of years in the presence of these poisons and have adapted to handle them in small concentrations. Thus, the objective is not to eliminate poisons altogether, which is impossible, but to hold concentrations to safe levels. As has been aptly said, "There are no safe substances, only safe amounts."[37]

According to EPA assessments, 27 percent of U.S. streams and shoreline miles were found to be polluted in 1970, and 29 percent in 1971.[38] The apparent 2 percent worsening was held to be primarily the result of improved field reporting, but gains in some areas seemed to be offset by losses in others.

Huge programs aimed at cleaning up waterways have produced notable gains in past decades and offer hope for the future, if we have enough determination. Two such programs are worthy of mention, one in the United States, and one in West Germany.

In the 1930s, the public became aware that the Ohio River of which

---

[35] *Death of the Sweet Waters, op. cit.*, pp. 154–160.

[36] W. S. Rukeyser, "Fact and Foam in the Row Over Phosphates," *Fortune*, January 1972, p. 71ff.

[37] "Natural Foods Fallacies," *U.C. News*, June 12, 1973.

[38] *Environmental Quality, op. cit.*, p. 11.

they were once proud had become a polluted mess. The pollutants of the thirties were the same ones we know now: sewage, chemicals, garbage, oil, acids, and so on. The subsequent cleanup of the Ohio has been a spectacular achievement, even though it is not yet complete. The problem was especially difficult to handle politically, because the Ohio passes through six states and has tributaries that originate in two more. The states pooled their interests in the formation of the Ohio River Valley Water Sanitation Commission (ORSANCO) in 1948. At the time of ORSANCO's formation, untreated municipal sewage and industrial waste were dumped into the Ohio with little thought. But, by the mid 1960s, 99.4 percent of the sewage was being processed through treatment plants, and 90 percent of the industrial plants had been brought under control standards. Acid discharges had been reduced, and other wastes, such as oil, were being disposed of in mile-deep holes instead of into the river. In the first 18 years of ORSANCO's operation, more than $1 billion was spent for municipal treatment facilities, and $500 million for industrial treatment.[39]

In West Germany, the Ruhr Basin is the most heavily industrialized area in Europe. It is served with a supply of water which is only a fraction of what would be considered necessary in the United States. The combined flow of the three major rivers draining the Ruhr Basin is less than that of the Delaware River at Trenton, New Jersey. The management of this vital water has been achieved through organization of the water users into large cooperative associations with far-reaching authority. The two largest rivers of the Ruhr have been maintained in a condition suitable for recreation and for municipal and industrial water supply. One of the smaller streams in the basin has been fully lined with concrete and is used exclusively for wastes. Near the point where this stream discharges into the Rhine, the entire flow is treated at once.[40]

Pesticides

Public concern over pesticides swelled rapidly following the publication of Rachel Carson's famous book *Silent Spring* in 1962. One of the main subjects of Miss Carson's book was DDT. Ten years later, after extended public debate and controversy, the Environmental Protection Agency banned nearly all uses of DDT in the United States.[41] The dispute still continued after this action, however. Proponents for the

---

[39] L. Edson, "The Rebirth of a River," in *Controlling Pollution, op. cit.*, p. 143.

[40] A. V. Kneese, "Water Quality Management by Regional Authorities in the Ruhr Area," in *Controlling Pollution, op. cit.*, p. 109.

[41] *Environmental Quality, op. cit.*, p. 125.

continued use of DDT maintained that we know more about DDT—both about its use and its behavior—than any other pesticide. Some of the substitutes which would have to be used, they pointed out, are highly toxic and could have even more detrimental long-range effects than DDT. Opponents to DDT maintained that its high persistence in the environment and its tendency to be concentrated through the food chain made the risks of continued use unacceptable.

DDT is only one of some 900 registered pesticides; in 1970 nearly 1 billion pounds of these chemicals, including insecticides, herbicides, fungicides, and rodenticides, were used in the United States.[42] In that year, the production of chlorinated hydrocarbons, the group of chemicals to which DDT belongs, accounted for 31 million pounds. This was down from the 1956 high of 244 million pounds. During the same time period production of the organophosphate pesticides, of which the well-known parathion is a member, rose from 7 million pounds to 57 million pounds. These figures show an important trend, which is the decline in the use of the more persistent but less toxic chlorinated hydrocarbons and the substitution of the less persistent but more toxic organophosphates.[43]

In the case of DDT especially, an enormous amount of research has been carried out to determine its toxicity to fish, birds, and mammals. Of special interest are experiments with human subjects. Volunteers have ingested large amounts of DDT, both by breathing it in mists and by eating it. Massive doses, on the order of 750 to 1000 mg., have been found to cause mild cases of illness. In long-term tests, volunteers have been fed up to 35 mg. of DDT per day (about 200 times the normal daily intake) for 21 months with no evidence of adverse effects.[44] On the other hand, thousands of cases of illness or death have been caused by pesticides of the organophosphate group. In California alone, 36 occupational fatalities among agricultural workers due to agricultural chemicals were reported during the period 1951–1970. Organophosphate pesticides accounted for nearly half of these.[45]

Even though DDT may not be especially toxic to humans, the same cannot be said for its effects upon fish and birds. Many accidental fish and bird kills have followed the use of DDT and other pesticides. DDT is obviously toxic to insects, but it is also toxic to other members of the phylum to which insects belong, a group known as the arthropods.

---

[42] D. Pimentel, *Ecological Effects of Pesticides on Non-Target Species* (Washington, D.C.: Office of Science and Technology, 1971).

[43] *Environmental Quality, op. cit.*, p. 17.

[44] *Report of the Secretary's Commission on Pesticides and Their Relationship to Environmental Health* (Washington, D.C.: U.S. Dept. of Health, Education, and Welfare, 1969,) pp. 295–300, 349.

[45] *Occupational Disease in California Attributed to Pesticides and Other Agricultural Chemicals* (Berkeley: State of California, Dept. of Public Health, 1970), p. 11.

Included in this group are creatures desirable for food, such as crabs, shrimp, and lobsters. Furthermore, DDT's persistence in the environment permits it to be accumulated through the food chain (from insects to fish to pelicans, for example) until it may reach high levels of concentration in the final host. It is known through laboratory studies that DDT can cause some kinds of birds (primarily predacious birds such as hawks, gulls, and pelicans) to produce eggs with eggshells 10 to 30 percent thinner than normal.[46] Whether thin eggshells occurring in a natural setting can be positively blamed on DDT is another matter. It has been widely assumed that nesting failures among southern California's brown pelican colonies were the fault of DDT because numerous broken thin-shelled eggs were observed in the nests. However, there are other known causes for eggshell thinning, such as when birds become nervous from too much commotion, motor boats, and the like.[47] At Anacapa Island, which is the principal pelican nesting site where thin eggshells were observed, the greatest commotion the birds had been subjected to in years was the repeated helicopter landings which occurred during investigations of the nesting sites. In 1971, one year after the demise of the brown pelican was announced, thousands of brown pelicans were nesting successfully on Anacapa and other islands.[48]

Another important matter for concern with regard to pesticides is whether some of the possible long-term effects might be to cause cancer (carcinogenesis) or to cause malformations in newborn infants (teratogenesis). No evidence exists of carcinogenic or teratogenic effects of pesticides in humans, but the fact that these effects have been observed in experimental mammals has caused widespread concern.[49] A concern of a different sort is that some experimental procedures on the carcinogenic effects of pesticides on mice involve force-fed doses 30,000 to 80,000 times the amount a person might take in per day on a weight-per-weight basis.[50] Whether such tests are meaningful to man may be open to dispute.

At one point in the debate leading to the DDT ban, alarm was expressed by some that the accumulation of DDT in the oceans would interfere with the ability of certain kinds of plankton to convert carbon dioxide into oxygen, thereby threatening the world's oxygen supply.

---

[46] *Ecological Effects of Pesticides on Non-Target Species, op. cit.*, pp. 177–178.

[47] *Agricultural Chemicals—Harmony or Discord for Food, People, and the Environment* (Berkeley: University of California Division of Agricultural Sciences, 1971), p. 75.

[48] R. G. Beatty, *The DDT Myth: Triumph of the Amateurs* (New York: The John Day Company, 1973), pp. 82–89.

[49] *Report of the Secretary's Commission on Pesticides and Their Relationship to Environmetnal Health, op. cit.*, p. 36.

[50] E. M. Mrak, *Speech Given at Environmental Toxicology Seminar,* University of California, Davis, February 4, 1970.

However, it was shown later that the level of concentration necessary to cause significant interference was far greater than could be produced in the oceans.[51]

The benefits of pesticides are beyond dispute, and we could not abandon them without worldwide catastrophe. Man's deadliest enemies in the world are the insects. It is believed that, until the nineteenth century, half the world's deaths were caused by insects in one way or another. In addition, insects and other pests compete with man for food. It has been estimated that without adequate pest control, food production could decrease by 30 percent.[52] In the United States this might mean only higher food prices and poorer quality, but in many places in the world it would mean famine since much of the world already lives on the edge of starvation.

The biggest killer insect of all has been the flea, because it is the usual vector of the plague. In the middle of the fourteenth century, plague killed an estimated 75 million people. In subsequent centuries it has erupted repeatedly. Even today, plague lurks just beneath the surface; several outbreaks have occurred in the United States in this century.[53] To control plague, man must control the rats which carry the fleas which carry the plague. Rodenticides are the chief weapon in this battle, together with cleanup of garbage and fifth which attract the rats.

Other insect killers abound. Mosquitoes have caused millions of deaths through malaria and yellow fever. The tsetse fly carries sleeping sickness and has killed hundreds of thousands of people in Africa. The "kissing bug" in Latin America transmits Chagas' disease, which kills one in ten of its victims. The louse carries typhus and has been responsible for the death of millions. During World War I, it is believed more soldiers in the French Army were killed by typhus than by bullets.

Malaria is one of the biggest killers and remains a critical problem in the world today. Prior to the discovery that the *Anopheles* mosquito was the carrier of the disease, malaria was blamed on bad air, especially marsh gas. After the discovery of DDT in the 1940s, it quickly became the principal weapon against the mosquito and still is in most of the world. The World Health Organization (WHO) credits DDT with having saved 5 million lives and preventing 100 million illnesses in the first eight years of its use. The annual malarial death rate in India has been reduced from 750,000 to 1500 through use of DDT. WHO states, ". . . this compound still provides the most effective, cheapest and safest means of abating and eradicating malaria . . . [and it has] served

---

[51] *Man's Impact on the Global Environment, op. cit.*, p. 25.

[52] K. Walker, "Benefits of Pesticides in Food Production," in *The Biological Impact of Pesticides in the Environment* (Corvallis, Ore.: Environmental Health Sciences Center, Oregon State University, 1970), p. 149ff.

[53] J. Clarke, *Man Is the Prey* (New York: Pocket Books, 1970).

2 billion people in the world without causing the loss of a single life by poisoning from DDT alone." WHO adds that, in view of this amazing record but with knowledge of the problems associated with increasing prevalence of DDT in the environment, its attitude toward DDT is "agonizingly ambivalent."[54] The appropriate use of DDT in the world thus presents us with a moral dilemma of the first magnitude. Do we continue using it, in the knowledge we are killing birds and perhaps mortgaging our own future? Or do we stop using it, in the sure knowledge that millions of humans will die as a result?

Many things have been done in the attempts to improve the use of pesticides and reduce their unwanted effects. Natural biological controls have been used where they are available; strict control procedures for pesticide use have been adopted; more effective and controllable methods for application have been developed. Constant efforts are exerted to reduce the use of all kinds of agricultural chemicals to the level of essential need. But it is clear that we will have to continue to live with pesticides. We will have a continuing struggle in the attempt to find the right balance for their use.

Mercury

General public alarm has been raised over the toxic effects of mercury, and with good cause, for mercury is very bad stuff. Organic compounds of mercury—especially methyl mercury—are even more toxic than elemental mercury. These poisons act directly on the brain, causing mental damage, paralysis, and death.

It has been known for centuries that mercury is poisonous, but the first major mercury disaster involving industrial pollution occurred in Japan, in 1953, at Minamata Bay. The final toll involved 397 cases of illness, with 68 deaths. The mercury came from a large chemical plant on the bay, was concentrated in the local fish populations, and the people were poisoned when they ate the fish. In 1973, the chemical company was ordered to pay $3.6 million in damages to the families who had been injured by the pollution.[55]

No comparable disasters have occurred in the United States, although there have been warnings issued concerning excessively high mercury concentrations in fish in the Great Lakes region. Not only that, but in 1970 it was discovered that canned tuna fish in the United States exceeded the government's guideline for mercury concentration of 0.5 ppm. Over 12 million cans of tuna were ordered withdrawn from the market.

---

[54] A. W. A. Brown, "The Present Place of DDT in World Operations for Public Health," in *The Biological Impact of Pesticides in the Environment, op. cit.*, p. 197.

[55] *San Francisco Chronicle,* March 21, 1973, p. 30.

Swordfish was also found to be in excess of the guidelines. Then it was discovered that the mercury concentration in preserved tuna specimens nearly 100 years old was greater than that in some recently caught specimens. The concentration in the preserved specimens averaged 0.95 ppm, nearly twice the allowable limit of 0.5 ppm. It was apparent, then, that mercury contamination in fish caught at sea could not be blamed on industrial pollution, but must have come from natural sources.[56] Also, we are forced to conclude that we have been eating mercury-contaminated tuna throughout this century.

Be that as it may, we can take no comfort from the fact that the oceans are contaminated by mercury from natural sources, or that there have been no proven deaths from mercury poisoning in the United States from eating fish. Mercury is still a poison, and there is general agreement that industrial discharges of mercury must be held to the lowest possible level—preferably zero. In 1970, a survey showed that a million pounds of mercury were being discharged into the U.S. environment from chlor-alkali plants alone. A subsequent survey of 50 varied industries showed that they had been able to reduce their mercury discharges by 86 percent.

Accidental mercury poisonings have occurred from eating seed wheat treated with methyl mercury fungicide. The most catastrophic such case occurred in Iraq in 1972, producing 6530 cases of poisoning and 459 known deaths. These poisonings occurred in spite of the fact that the grain was colored with a brownish-red dye and that the sacks had written and diagrammatic warnings. Unfortunately, the warnings were in a language unfamiliar to the populace. However, it is obvious that many of those poisoned knew they were dealing with poisoned grain, because they washed the grain before eating it—thus removing the reddish dye—and apparently believed they had also removed the poison.[57]

Oil spills

In 1971, the Coast Guard reported that there were 8496 spills of polluting substances in U.S. waters, mostly involving oil. The great majority of these spills were small, but many of them involved large-scale cleanup and disposal actions. About 7600 of the spills occurred in inland waters, bays, or harbors, with the remainder occurring in the ocean.[58] Throughout the world, an average of 15 major ship collisions

---

[56] J. J. Putnam, "Quicksilver and Slow Death," *National Geographic Magazine*, October 1972, p. 507ff.

[57] F. Bakir *et al.*, "Methylmercury Poisoning in Iraq," *Science*, July 20, 1973, p. 230ff.

[58] *Environmental Quality, op. cit.*, pp. 118–120.

or oil well blowouts occurred each year from 1969 through 1971, resulting in oil spills.[59]

Probably nothing has aroused more environmental outrage than the sight of a dying bird soaked with oil. Hundreds of volunteers have rallied to the scenes of major oil spills to try to save the birds. Unfortunately, their efforts have met with little success. In the case of the 1969 Santa Barbara oil spill, for example, it was claimed that only 10 to 15 birds out of 1500 treated for oil damage ultimately survived.[60] Shoreline clean-up efforts have also been difficult and frustrating. In the case of the Santa Barbara spill, after an immense clean-up operation involving 1000 men and 10,000 truckloads of straw, the final cleansing of the beaches was accomplished only by the natural scouring action of the tides and waves.[61]

Major oil spills which occur close to shore almost always result in killing shorebirds, with the deaths sometimes numbering in the thousands. Losses like these probably make only a small dent in the bird populations; after all, natural diseases such as botulism kill millions of birds each year. Nevertheless, wild birds are very precious to man, not only on economic grounds. A world without birds is unthinkable to most people. Hence, bird kills must be regarded as one of the principal adverse effects of an oil spill.

Two of the world's best-known oil spills are those of the *Torrey Canyon*, involving a tanker which was wrecked off Great Britain in 1967, and the Santa Barbara oil-well blowout in 1969. In the case of both of these spills, which occurred in fairly deep water, the actual biological damage from the oil itself turned out to be relatively slight, except for the tragic destruction of sea birds.

The *Torrey Canyon* disaster involved 700,000 barrels of crude oil, nearly 10 times the amount at Santa Barbara. A report by the Marine Biological Association of the United Kingdom concluded that the worst effects came from the detergents which were used to disperse the oil. The detergents, which killed countless shellfish and other animals, were employed because, in the Association's words, ". . . the preservation of coastal recreational amenities was of first priority. . . ." The crude oil itself was unsightly and stank, but was relatively benign biologically except to the birds. Limpets and barnacles survived, even though coated with oil. The Association noted, "No actual toxic effects of the oil have been noted either in shore or pelagic [deep-water] animals."[62]

---

[59] *The Pulse of the Planet, op. cit.*, p. 118.

[60] *Ibid.*, p. 27.

[61] C. Steinhart and J. Steinhart, *Blowout; A Case Study of the Santa Barbara Oil Spill* (Belmont, Calif.: Duxbury Press, 1972), p. 100.

[62] *"Torrey Canyon" Pollution and Marine Life: A Report by the Plymouth Laboratory of the Marine Biological Association of the United Kingdom* (Cambridge, England: The University Press, 1968), pp. 14, 178.

In the Santa Barbara case, assessments after the spill were similar to those of the *Torrey Canyon*: little lasting biological damage except to birds. However, controversy continues to exist over the possibilities of subtle undetected effects on planktonic life or on bottom-dwelling organisms, both of which are important in the marine food chain.[63] One of the problems besetting the assessment of biological damage was the fact that the oil spill had coincided with unusually heavy winter storms which had caused 14,000 gallons of silt to enter the Santa Barbara Channel for each gallon of oil. Since fresh water is deadly to many kinds of marine life, it was difficult to separate damage done by the flooding from that done by the oil. Likewise, controversy raged over whether sea lions and whales had been damaged. Some dead sea lion pups were observed covered with oil, but it could not be determined if the oil was the cause of death or not. Press reports stated that six dead whales had washed ashore on the coasts of California, and the assumption was that they were casualties of the spill. But it developed that one had died before the spill, one had been harpooned, the death of another was caused by pneumonia, and autopsies on the remaining three showed no traces of oil.[64]

Another complicating factor at Santa Barbara is that natural oil slicks have been reported in the channel ever since white explorers first arrived. Residents of the area have long known that it was impossible to go to the beach without getting gobs of oil on one's feet. (Author's note: This is from direct personal experience, for I was once a resident there.) Deposits of oil lie very close to the surface in the Santa Barbara Channel, and natural leaks of oil coming from the ocean floor have long been known. Perhaps this is one of the reasons so little biological damage was reported as a result of the 1969 spill: the species in the channel had been living in the presence of oil and were able to cope with it.

An important question in oil spill cases is: Could they have been prevented? In both the *Torrey Canyon* and Santa Barbara cases, the answer would appear to be yes. It is claimed that the *Torrey Canyon* ran aground on Pollard Rock in trying to save half an hour by taking a treacherous route past Seven Stones Reef.[65] At Santa Barbara, the blowout appeared to have been caused because the crew was pulling out the drill string too fast. This upset the hydrostatic pressure balance at the bottom of the hole, which caused gas and oil to come surging out the top. In the past, such an event was called a "gusher" and was usually regarded as a sign of oil prosperity. The largest gusher on record occurred in 1910 near Taft, California, and spewed out 9 million

---

[63] R. Easton, *Black Tide: The Santa Barbara Oil Spill and Its Consequences* (New York: Delacorte Press, 1972), pp. 254–256.

[64] *Blowout, op. cit.*, pp. 81, 83, 97, 101.

[65] *Ibid.*, p. 127.

barrels of oil—130 times as much as the Santa Barbara spill—before it could be stopped. Emergency blowout preventers have been developed, and one was used at Santa Barbara. The preventer shut off the blowout at the top, but the pressure forced the oil out through weak rock strata lying just beneath the ocean floor, because the well had been protected with steel casing to a depth of only 239 feet. Since drilling at the time had extended to 3479 feet, usual practice would have been to protect the well with casing to a much deeper level than 239 feet. A waiver of the rules had been granted by the U.S. Geological Survey, because it was felt the subsurface structure was too weak below the 239-foot level to cement the casing safely in place. However, other experts contradicted this view, and stated the casing could have been cemented safely at 1200 or 1800 feet. If this had been done, the blowout might have been stopped by the emergency prevention device.

In justifying their actions, the drillers pointed out that four other wells had been completed safely just ahead of the one that blew out, from the same platform, in the same rock formation, and using the same procedures. Furthermore, this was the first time such an accident had occurred off California, even though a thousand production wells had previously been drilled in the state's waters.[66]

The Santa Barbara spill precipitated a highly charged political battle over the future of offshore drilling for oil. The residents of Santa Barbara had been fighting the encroachment of oil for years. It was bad enough to look out on their scenic channel and see it sprouting 20-story oil derricks. But to have the oil break loose and surge ashore to foul their beaches and harbors was the last straw. The fact that Union Oil Co. had to spend more than $10 million in cleaning up the spill satisfied no one. Large numbers of people across the country joined in the fight against oil. Many opposed further oil exploration in the United States, but apparently favored it in foreign countries because they urged that the United States rely more on importation. Others opposed increased importation because this meant more tankers and maybe more oil spills. Still others urged the cessation of offshore drilling and urged exploitation of oil shale reserves in the state of Colorado instead. They overlooked the fact that oil shale exploitation implied the need for surface mining on a massive scale.[67] Others simply cried out that because of our civilization's reliance on oil we were creating a dead sea off southern California, upsetting the earth's oxygen balance and causing the earth's temperature to rise, which would melt the polar ice caps.[68]

The biological damage from oil in the *Torrey Canyon* and Santa Barbara spills may have been relatively slight, but there have been

---

[66] *Black Tide*, *op. cit.*, pp. 123, 182, 202–203, 206.
[67] *Ibid.*, pp. 126–127.
[68] *Ibid.*, pp. 124, 126–127, 191.

other spills which have been much more lethal. Open-water spills involving crude oil generally cause little damage—not even to seabirds, unless the oil comes ashore. In 1971, for example, 17 major spills were included in the Smithsonian Institution's report on short-lived phenomena; in 10 of these, the biological damage was described as moderate to light, in three of them the damage was heavy, and in four others the damage was not known.[69] However, spills which occur in harbors and estuaries are potentially much more damaging than open-water spills, because there is less opportunity for natural dissipation and biodegradation. (Yes, there are bacteria which "eat" oil, causing natural biological degradation—a process which may require several months.[70]) Also, refined products, such as diesel oil, are much more lethal than crude oil. For example, in the 1969 Buzzards Bay spill in Massachusetts, massive numbers of fish, lobsters, shrimps, crabs, and scallops were killed by a spillage of 175,000 gallons of diesel oil. Surprisingly, few shorebirds were harmed by this spill.[71]

Many schemes have been devised for containing and cleaning up oil spills. After the Santa Barbara spill, 14 major oil companies joined together to form Clean Seas Incorporated, whose purpose is to provide an emergency force to cope with oil spills.[72] As of 1973, 83 similar cooperatives had been organized nationwide. Some of the spill control techniques under development were floating barriers to contain spills; surface film-forming chemicals to prevent spreading; absorbent foams to aid in recovery; and microbial cultures to speed biological degradation.[73]

Recycling and solid wastes

It has become popular to depict America as ". . . a nation knee-deep in garbage, firing rockets to the moon."[74] Such a dramatic image conveys in one package all the tragedy of an alleged "plastic society" wasting its talents while it drowns in its own muck.

While grossly exaggerated, the notion of an American buried in garbage does expose some serious problems. For example, the packaging industry of America consumes 55 million tons of materials per year, most of which winds up being burned, dumped, or buried. Historically, open dumping or burning were the preferred disposal methods, but controls on air pollution and on visual eyesores are eliminating these

---

[69] *Pulse of the Planet, op. cit.*, pp. 88–112.

[70] *"Torrey Canyon" Pollution and Marine Life, op. cit.*, pp. 11–12, 81–85.

[71] *Pulse of the Planet, op. cit.*, p. 48.

[72] *Black Tide, op. cit.*, p. 217.

[73] "Cleaning Up Oil Spills Isn't Simple," *Environmental Science and Technology*, May 1973, pp. 398–400.

[74] Garrett de Bell (Ed.), *The Environmental Handbook* (New York: Ballantine Books, 1970), p. 214.

as viable alternatives. Even the method of burying in sanitary landfills is running into trouble in some urban areas because suitable sites are hard to come by.[75] Also, many environmental broadsides have been aimed at the increasing fraction of solid waste consisting of plastics, because most plastics resist biodegradation. Disposable cans and bottles litter the countryside and waste our natural resources. The manufacture of the aluminum can consumes six to eight times as much energy as does steel.[76]

Attempts to recycle bottles, cans, and newspapers have mostly been marked with frustration. Recyclable bottles are heavier than throw-aways and cost more. Bottlers need to get six trips per bottle to break even, but in some urban areas the average is only four. Consumers apparently prefer to throw away their containers, even though it costs more to do so.[77] Volunteer recycling efforts have not succeeded because the market structure is not adequate to meet costs and sustain the manpower needed to do the job. Recycling, to be successful, must be institutionalized. Manpower costs must be met through sale of recycled material, taxation, or a combination of the two. However, there is little doubt that recycling will have to enter our lives in a big way. The increasing scarcity of high-grade raw materials leaves no choice in the matter. As we are forced to utilize low-grade ores, recycled material will look more and more attractive, just from an economic viewpoint.

Metals already are recycled to a considerable degree. According to the U.S. Bureau of Mines, 50 percent of the country's lead, 45 percent of the iron and steel, 40 percent of the copper, and 25 percent of the zinc and aluminum used each year are obtained by recycling. The Bureau takes the view that dumps and junkyards represent "man-made mines" containing essential resources. It sponsors scores of research projects aimed at economic recovery of waste materials.[78]

An interesting pilot plant experiment in recycling solid waste is that at Franklin, Ohio. In this plant, unsorted refuse is fed into a pulping machine. The metal fragments are screened out, and ferrous materials are separated from nonferrous matter magnetically. Glass and sand are separated by centrifugal action. Final screening reclaims usable paper fibers. The metals, the glass fragments, and the paper fibers all have market value and can be sold. The remainder, consisting of plastics and other organic material, is burned. It has been proposed that the heat from burning could be used to generate electrical power. Another proposal is that organic wastes be converted into oil. If all the readily collected organic residue of the United States in 1971 were

---

[75] *Environmental Quality*, *op. cit.*, p. 204.

[76] E. Faltermayer, "Metals: The Warning Signals Are Up," *Fortune*, October 1972, p. 109ff.

[77] T. Alexander, "The Packaging Problem Is a Can of Worms," *Fortune*, June 1972, p. 105ff.

[78] C. B. Kenahan, "Solid Waste: Resources Out of Place," *Environmental Science and Technology*, July 1971, p. 594ff.

to be converted, it could make 170 million barrels of oil, enough to satify 3 percent of our annual oil needs. Oil conversion pilot plants have been constructed by the U.S. Bureau of Mines in Pittsburgh, and by Garrett Research and Development Company near San Diego, California.[79] With larger, improved plants of the future, and in view of growing scarcities of materials, solid-waste handling plants are expected to become money-makers.[80] Other possibilities exist for recycling organic wastes. In a suitably designed furnace, for example, solid wastes could be converted into a gas consisting mostly of CO and H_2, which could then be treated further to create methanol (methyl alcohol, CH_3OH).[81]

It is a curious fact that the use of much of the packaging material which turns up as solid waste may actually prevent other kinds of waste. It has been estimated that modern packaging in some cases prevents as much as 20 percent of the food spoilage which would otherwise occur.

Main target: the automobile

To enumerate all the ways in which the automobile has come under environmental attack would be a considerable task. First and foremost, of course, the auto has been identified as the major polluter of urban air. Other indictments are that it consumes excessive amounts of energy and raw materials; causes enormous amounts of land to be locked up in freeways, parking lots, and gas stations; has generated a second-home boom that has destroyed attractive rural areas; has created massive traffic jams in national parks; is a major killer—responsible for some 50,000 deaths in the United States per year.

Criticism of the automobile has assumed a tone of strong moral censure. One critic describes our love affair with the automobile as an "embrace of death." Drivers, he states, hurtle along freeways, ". . . bent tensely over their steering wheels with riveted, lusterless eyes that see nothing but pavement ahead. . . ." Another critic charges the automobile with having insulated man both from the environment in which he lives, as well as from contact with other humans.[82] Further, it is alleged that Detroit has pandered to the baser instincts of the masses by offering them chrome-bedecked, overpowered monsters that serve mostly as status symbols.

---

[79] A. L. Hammond, W. D. Metz, and T. H. Maugh, *Energy and the Future* (Washington, D.C.: American Association for the Advancement of Science, 1973), pp. 73–78.

[80] "Reclaiming Municipal Garbage," *Environmental Science and Technology*, October 1971, pp. 998–999.

[81] T. B. Reed and R. M. Lerner, "Methanol: A Versatile Fuel for Immediate Use," *Science*, December 28, 1973, pp. 1299–1304.

[82] *The Environmental Handbook, op. cit.*, pp. 120, 197ff.

Some people have urged that we simply do away with the auto. Instead, we should walk, ride bicycles, use mass transit, or just stay home. But such an extreme view overlooks some basic reasons why Americans —and nearly everybody else in the world, for that matter—love their automobiles. For only a small percentage is any "love affair" with chrome, horsepower, or prestige involved. Most people love their cars because of the benefits they offer. Cars take people to the beach, to the mountains, or to rivers and lakes for weekends and holidays. They permit families to travel and to see the country in which they live. They make it possible for relatives and friends to visit, who might otherwise make contact seldom or never. They make possible errands to the store, to the doctor, to school, and to make business contacts that otherwise would be difficult or impossible. A very revealing result of a study conducted in 1962 by the Outdoor Recreation Resources Review Commission was that "driving for pleasure" outranked all other forms of outdoor recreation in the United States.[83]

The struggle of the auto industry to do something about air pollution is a familiar story. Since the early 1960s, one pollution control device after another has been added to the internal combustion engine (or, more properly, to the "spark-ignition engine"), coupled with numerous engine redesigns. Crankcase emissions of HC have been eliminated by feeding them back through the combustion process. Evaporative emissions from carburetor and fuel tank have been brought under control. Leaner air fuel mixtures have been utilized to reduce the production of HC and CO. Exhaust reactors of various types have been developed to oxidize any remaining HC and CO that may be produced. Exhaust gas recirculation (EGR) systems have been developed which feed some of the exhaust back into the cylinder, reducing the peak combustion temperature and thus inhibiting the formation of NO_x.[84] As a result of these and other measures, from 1961 to 1970, car emissions of HC had been reduced 80 percent, CO by 70 percent, and NO_x by 40 percent.[85] Just one company—General Motors—calculated it had spent $350 million on emission controls in 1973, including research, development, new facilities, and tooling. Its investment from 1967 to 1973 had exceeded $1 billion, and it estimated it would spend another billion from 1974 to 1976.[86]

"Why," ask some, "do we remain wedded to the spark-ignition

---

[83] *Outdoor Recreation for American, op. cit.*, p. 4.

[84] *Project Clean Air*, Vol. I (Riverside, Calif.: University of California, September 1, 1970), pp. 2–14ff.

[85] *Automotive Powerplant Research.* Statement by W. G. Agnew, presented to Senate Subcommittee on the Environment and Senate Subcommittee on Science, Technology and Commerce, Washington, D.C., May 14, 1973.

[86] Statement by Edward N. Cole, President, General Motors Corporation, presented to the Subcommittee on Air and Water Pollution of the Senate Committee on Public Works, Washington, D.C., May 30, 1973.

engine? Surely there must be something else." The answer is that there has been nothing else available to compete successfully with the spark-ignition engine in matters such as power/weight ratio, fuel economy, performance, and durability. Seventy years and billions of dollars have gone into the development of the modern auto engine, and one does not easily overcome a headstart like that. Nevertheless, the auto companies are trying. For example, General Motors claimed it had spent $36 million during 1972 and another $46 million in 1973 in the search for alternate power plants.

Steam engines and electric cars have been two of the public's perennial favorites as alternates to the gasoline engine. But both of these lost out to the gasoline engine early in this century and have as of this date never overcome their handicaps. To the early auto experimenters, the choice was easy to make. Not only did the electric car suffer from an extremely limited range, and the steam car from the constant danger of explosions, but both of these suffered a severe weight disadvantage. For similar power output, a steam power plant weighed twice as much as a gasoline engine, and an electric power plant weighed nearly five times as much.[87] In the 1960s, General Motors and others were carrying on active research programs on steam cars, but by 1973, GM had dropped out, citing excessive weight, poor fuel economy, cost, and other problems as too formidable to overcome.

Electric cars, operated by batteries, seem to many people to be an ideal nonpolluting alternative to gasoline-powered cars, but this is of course a delusion. The electricity needed by an electric car has to be generated somewhere and is generally created by burning oil or coal in a central power plant, although it can be argued that central power plants can successfully be subjected to more stringent pollution controls than can individual autos. Furthermore, the heavy weight penalty and low effective range of electric cars still have not been overcome. They will find utility in certain special ways but do not appear in the foreseeable future to be a viable broad-scale alternative to the gasoline-powered car.

Other alternatives are of active interest to the auto companies, and some were entering the market in the 1970s.[88] Prominent among these was the rotary engine, which offered the advantage that it was only half the size of a comparable piston engine. Another entrant was Honda's "stratified-charge" engine, which promised good fuel economy and low emissions. In a stratified-charge engine, two carburetion systems are employed, one of which injects an extremely lean mixture (too lean to be ignited by the spark plug) into the main cylinder, while the other places a rich mixture in a small region near the spark plug.

---

[87] D. L. Cohn, *Combustion on Wheels* (Boston: Houghton Mifflin, 1944), p. 40.

[88] See *Automotive Powerplant Research, op. cit.*

The plug ignites the rich mixture, which in turn ignites the lean mixture. The overall effect is to promote good combustion and low emissions. Both of these novel engines, at the time of their introduction, had only been developed for small cars.

Gas turbines and diesel engines have been much discussed as alternative power plants for autos. Both have low emissions of HC and CO, but the gas turbine had a problem of NO_x emissions, and the diesel engine is given to emitting clouds of smelly black smoke if not properly maintained. Contrary to popular impression, diesel engines do not contribute much to smog, but this is because their numbers are not significant when compared to autos. Both gas turbines and diesel engines have a weight disadvantage for auto applications, which makes them more useful for trucks and buses than for autos. It is true that diesels have been used in European cars, but GM claims that if one of these were to be scaled up to the size necessary to power an American auto, it would be difficult to fit it into the engine compartment.[89]

Other possible alternatives to current engines are being explored, such as hybrid electric/gasoline engines and the use of gaseous fuels such as liquefied petroleum gas (LPG), natural gas, and hydrogen. Methanol (methyl alcohol, CH_3OH) and ethanol (ethyl alcohol, C_2H_5OH) have also been proposed as alternate fuels. The latter have an advantage over the gaseous fuels because they are liquid and so can be carried readily by an auto without using pressurized tanks. All of the foregoing fuels are inherently less polluting than gasoline. Methanol appears more attractive than ethanol, because it can readily be made from a fossil fuel such as coal, whereas ethanol is usually produced by fermentation of grains and would thus be competitive with our needs for food.

There is one other alternative power plant which deserves mention: the fuel cell. Although the basic discovery of the fuel cell principle was made in 1839, by the 1970s fuel cells had not yet significantly entered the commercial scene. However, an enormous investment in fuel cell development was made during the space program, and if the fuel cell offers us any advantages in our day-to-day lives, it will be another domestic benefit from our investment in space.

There are numerous kinds of fuel cells, involving different kinds of fuels and chemical reactions. The most basic, and one of great potential for practical use, is the one which uses hydrogen. Hydrogen and oxygen from the air are combined in the presence of an electrolyte and a catalyst to create an electric current. The process is essentially the reverse of electrolysis. The "pollution product" is water. Much interest has been aroused over the use of hydrogen as a fuel, because it can be generated from water, and when it is consumed, we get water back

---

[89] *1973 Report on Progress in Areas of Public Concern* (Warren, Mich.: General Motors Corp., February 8, 1973), p. 42.

again. There could hardly be a more attractive energy cycle. To be sure, there are problems. We need to develop compact storage methods for hydrogen to be able to carry enough in a car. It is explosive (but so is gasoline). Most important of all, we need to develop basic nonpolluting sources of energy, such as fusion power or solar energy, in order to make the hydrogen in the first place. In the meantime, much research and development has gone into fuel cells. *Science* magazine estimated that $50 million had already been invested in the potential commercial exploitation of fuel cells by 1972, mostly by U.S. companies, and that $100 million more would be invested by 1975. The objective was to achieve full-scale commercialization during the decade of the 1970s.[90]

As for other problems surrounding the auto, one can take note of certain promising trends. Certainly, during the decade of the 1960s autos had grown safer and would continue to grow safer in the 1970s. From 1967 to 1972, the highway fatality rate in the United States dropped from 5.6 per 100 million vehicle miles to 4.5—by far the lowest in the world.[91] True, the improvements in safety had been accomplished in spite of the skepticism of the auto industry and the outright hostility of the average consumer. But by the seventies, belt restraints, air cushion bags, and padded interiors were a reality—even though most drivers refused to use belt restraints even when they were present.[92]

Another encouraging trend in the 1970s was that smaller cars were growing more prevalent in the United States, because of growing pressures on fuel supplies and other resources. In the four-year period from 1969 to 1973, the share of the U.S. market taken by small cars had risen from 22 percent to 40 percent. Furthermore, 60 percent of the small cars being sold were manufactured in the United States, up from 40 percent in 1969.[93]

As for the crush of cars in cities, less optimism is justified. A government proposal in 1973 that gasoline be rationed, and other limits placed on the use of autos in cities, brought immediate angry protests coast to coast.[94] Clearly, the solutions for cars in cities will be difficult. More mass transit systems can help alleviate the problems but will not completely solve them. Perhaps some limits on urban auto travel will ultimately be necessary. Better planning for future cities may provide a partial answer, according to economist Wilfred Owen. Some cities already in existence around the world have developed ways in which people can live near their work in integrated communities, and thus reduce the need for commuting. According to Owen, these include the

---

[90] "Fuel Cells: Dispersed Generation of Electricity," *Science*, December 22, 1972, p. 1273ff.

[91] *Traffic Engineering*, January 1974, p. 53.

[92] *The Sacramento Bee*, October 28, 1973, p. B9.

[93] *Time*, July 2, 1973, p. 56.

[94] *Time*, June 25, 1973, p. 54.

new Pakistani capital, Islamabad; a new town in Scotland, Cumbernauld; and "perhaps" two planned cities in the United States, Reston, Virginia, and Columbia, Maryland.[95]

Population and resources

In 1968, Paul Ehrlich's *The Population Bomb* burst upon a startled public. Soon thereafter it became fashionable to speak in terms of "population doubling times" and the "population explosion." It was noted that, in the seventeenth century, the time necessary to produce a doubling of the world's population was 250 years, but by 1970 it had become only 33 years.[96] If population increases were to continue thus, the earth would soon clearly have "standing room only." In the United States, an especially worrisome omen of the 1960s and 1970s was the fact that the post-World War II "baby-boom" was just entering the childbearing years, which would cause an "echo-boom" and force the population to spurt ahead once more.

Surprisingly, in 1971, just about the time the "echo-boom" was supposed to force the birthrate upward, the opposite happened instead: the birthrate went down. Even though the decline was not great (from 18.2 births per 1000 population in 1970, to 17.3 in 1971), the news media immediately spread the notion that we had entered a period of "baby-bust" and "birth dearth." Many persons assumed from these figures that the American people, alarmed at the prospect of future overpopulation, had voluntarily put on the brakes and now everyone could breathe easier. The President's Commission on Population Growth and the American Future was not reassured, however. It warned:

> The baby-bust psychology may give rise to unwarranted complacency born of the notion that all of the problems associated with population growth are somehow behind us . . . Even if immigration from abroad ceased and couples had only two children on the average—just enough to replace themselves—our population would continue to grow for about 70 years. Our past rapid growth has given us so many young couples that, to bring population growth to an immediate halt, the birth-rate would have to drop by almost 50 percent, and today's young generation of parents would have to limit themselves to an average of about one child. That is just not going to happen.[97]

---

[95] E. T. Chase, "How to Save the Cities from the Cars," *Fortune*, September 1972, pp. 173–174.

[96] D. H. Meadows *et al.*, *The Limits to Growth* (New York: The New American Library, 1972), p. 41.

[97] *Population and the American Future, The Report of the Commission on Population Growth and the American Future* (New York: The New American Library, 1972), p. 15.

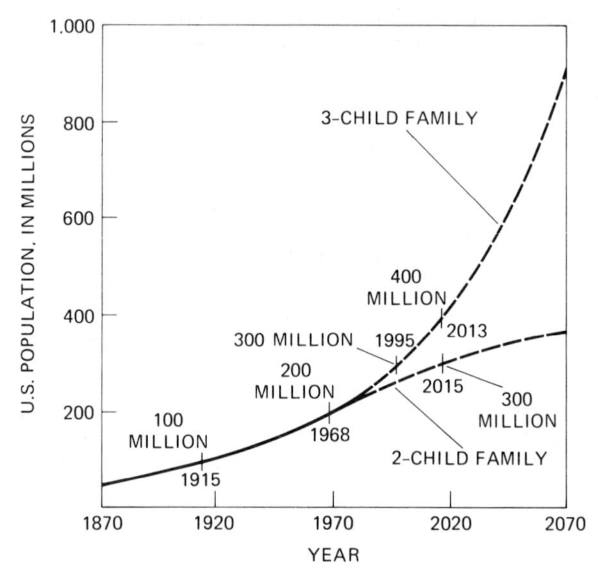

Fig. 1–2 Population projections for the United States, based upon two alternate assumptions: 1) that family size averages two children, and 2) that family size averages three children. (Source: *Population and the American Future.*)

By 1973, the U.S. birthrate had declined even further, to 14.8 births per 1000 population, but was still far from the 50 percent reduction required to halt our population growth altogether. Figure 1-2 shows the projections for the United States two ways: 1) if every couple limited themselves to two children each, and 2) if the average family consisted of three children. Obviously, the three-child family produces an uncontrolled explosion if continued, whereas the two-child family can gradually produce stability. Even so, the two-child family produces a population of 300 million people in the United States by 2015. Thus, well within the life span of those who were college students in the 1970s, the United States will have to learn to cope with 100 million additional people.

The implications of such population growth for pollution and resource consumption are obvious. Every one of those 100 million extra people will want his or her share of resources—food, water, air, energy, living space, recreation, schooling, and everything else. And everything which is consumed is necessarily rejected later as waste: each and every one of us is a polluter.

Concern about population has undergone several cycles in the United States in this century. Just before World War I, the country was warned that it was committing "race death" by not having enough children; then, by the 1920s, the pendulum had swung to the point where the public was afraid it would soon have too many people. In the depression of the 1930s, the pendulum swung again as birthrates fell to record lows and some experts predicted that by 1965 the population would start to decline.[98]

During the periods when birthrates have gone to low levels, voices have been heard to cry out that declining population growth would bring with it economic disaster. In fact, some economists have argued that the Great Depression of the 1930s occurred *because* the population was not growing. Japan, which had adopted a widely publicized policy of population control following World War II, seemed to be in the process of reversing itself in the 1960s and 1970s. In 1963, Prime Minister Ikeda proclaimed, ". . . when the population is increasing the nation is also prospering," and in 1970, Prime Minister Sato declared that it would be necessary in the future to restrict abortions in order to provide a sufficient labor force and insure national survival.[99]

Besides the effects of increased population upon pollution, there is a relentless demand exerted upon resources. The first resource to come under pressure is food. In the early 1970s, the United States was startled to discover how quickly its huge grain surpluses could disappear as a result of demand from abroad. The nation also was surprised to find that its food costs were forced upward because foreign countries were willing to pay high prices. Americans had become accustomed to paying only 16 percent of their disposable personal income for food, while in Germany the share going for food was 22.5 percent, and in Japan 33.2 percent.[100] It also became apparent how completely the world's food supply depends upon the weather: a few bad years in succession could produce disaster. Even the much-heralded "Green Revolution" had failed to produce the expected gains in world food supply, largely because of its requirements for fertilizers, irrigation, and acquisition of new skills.[101]

Other resources to come in short supply are energy, water, and metals. The United States began to feel its first real energy pinch in the 1970s, as it faced first a fuel oil shortage and then a gasoline shortage. America found it was going to be necessary to import a growing share of its energy resources from abroad, mostly in the form of oil, which put this country's desires directly into competition with the ever-expanding needs of Europe, Russia, and Japan.

---

[98] E. Pohlman (Ed.), *Population: A Clash of Prophets* (New York: The New American Library, 1973), pp. 2, 79.

[99] *Ibid.* pp. 70, 139–140.

[100] *Time*, April 9, 1973, p. 13.

[101] "World Food Situation: Pessimism Comes Back Into Vogue," *Science*, August 17, 1973, p. 634ff.

As for water, it probably would not seem to most Americans that there was likely to be any shortage, except perhaps in the southwest. Yet, in 1972 the Commission on Population Growth and the American Future warned that the area of water shortage in the United States would spread eastward and northward across the country in the next 50 years.[102] In Japan, water rationing was instituted in 1973, partly because of that country's huge industrial establishment, with its accompanying appetite for water.

With metals, the picture appears a little less bleak. The United States has large stores of minerals but imports nearly half of its raw material anyway, because so many foreign sources are cheaper. If prices were to rise appreciably, or as new technological developments arise, many of the low-grade U.S. deposits would appear more attractive. For example, even though the high-grade iron ore of the Mesabi Range in Minnesota is nearly gone, the United States has large deposits of taconite. New developments in technology have made it possible for two thirds of domestic iron ore needs to be met from this source. Abroad, vast new reserves of iron and aluminum ores have been discovered in Australia.

For most metals, the world supply appears ample for a century or so—a period which is beyond that with which human predictive ability can cope. But for seven important metals, the known world supplies will last for only a few decades, and then we will either have to find substitutes, recycle, or do without. These are zinc, cobalt, platinum, silver, mercury, tin, and tungsten.[103]

Concern over nonrenewable resources such as metals and fossil fuels makes us consider just what our *renewable* resources might be. Obviously, one way to make many resources renewable is to recycle them. Food is a renewable resource, and so is water, which is conveniently recycled for us by the solar engine. Timber is an exceedingly important renewable resource. Experiments are under way which may result in genetically superior trees which will grow twice as fast as at present.

Energy is our greatest concern. Only a tiny fraction of our present energy resources—that produced by falling water—can be regarded as renewable. All the rest, including nuclear energy, is obtained by mining the earth. Such reflections have caused thoughtful men to turn toward the greatest energy source of all—the sun. On a human time scale, the sun's energy is not only renewable but is limitless. The challenge is immense, but in time, all of man's needs will have to be met from naturally renewable sources or by recycling.

In 1968, a group of 30 individuals from ten different countries formed The Club of Rome and initiated an ambitious undertaking—the Project on the Predicament of Mankind. Phase One of the project

---

[102] *Population and the American Future, op. cit.*, pp. 60–61.

[103] "Metals: The Warning Signals Are Up," *Fortune*, October 1972, p. 109ff.

was undertaken by a group at MIT and involved the application of system dynamics to a computerized model of the world. The results were published as a book: *The Limits to Growth.* Under most of the conditions assumed in the various trial runs of the project, the computer predicted that disaster would overtake the world before the year 2100, and under some conditions well before that time. Disaster would either take the form of unlimited pollution, catastrophic population decline, or both. Only under circumstances of population stabilization at a fairly low level, coupled with resource recycling and strict pollution controls, did the computer predict a stable future for the human race.[104] One could perhaps argue with the details of system dynamics or with the assumptions of the global model. It has been shown, for example, that MIT's world model is very sensitive to changes in assumptions.[105] Nevertheless, one cannot seriously argue with the fundamental fact that there *are* limits to growth. If populations try to grow without limit, nature eventually sees to it that they level off—by increasing the death rate. The great challenge for civilization will be to find ways to make the smooth transition from an economy based on growth to one that is not.

Most projections of the future necessarily omit a vital factor—innovation. In making models of the future, we dare not include any unexplained developments which may turn out to have substantial control over the results. To do so would be to risk the ridicule of our scientific colleagues who know that one is not allowed to introduce such "miracles" into our projections. Yet, by looking into the past, we know very well that our fate has been most profoundly affected by those developments which were completely unforeseen.

In 1900, for example, one could not possibly have been able to foresee the tremendous importance the automobile would have in our lives, or the role of the manufacture and massive distribution of petroleum products. And, as recently as 1950, no one could have foretold that the auto would be such a storm center for environmental and political attack as it has become.

As another example, in 1937 the National Research Council issued its report on *Technological Trends and National Policy.* This report undoubtedly was prepared by as well-informed a scientific group as the nation had available, but it failed to foresee atomic energy, radar, antibiotics, or jet propulsion, all of which emerged in the next five years.[106]

More recent examples are just as striking. In the late 1950s it had become fashionable in the computer industry to predict that the market

[104] *The Limits to Growth, op. cit.,* Chap. V.

[105] R. Boyd, "World Dynamics: A Note," *Science,* August 11, 1972, pp. 516–519.

[106] *Environmental Quality, op. cit.,* p. 69.

for large-scale computers would soon be saturated—not many organizations could afford them, and there wasn't that much application for them anyway. Within the next 20 years, as is now well known, the computer industry had raced through several successive generations of computers and had grown into a giant enterprise. Also in the 1950s, one business machine company dropped its development project for a portable electronic calculator because so many transistors would be required that the unit would have to sell for $5000. Only 15 years later, transistors had been superseded by integrated circuits, and portable electronic calculators were available on the market for less than $100.

In the case of pollution, it is commonly assumed that environmental improvement will only be achieved through higher costs. Initially, this may be so, but we should not overlook the fact that pollution *causes* enormous costs, through poorer human health, crop damage, fish kills, materials decay, and the like. We should also not overlook one of the fundamental facts of our economic life ever since the industrial revolution: the foremost effects of the activities of engineers through the years have been to reduce costs and increase benefits. This is what engineering is all about. There is no reason why we should expect that engineers will suddenly fail to be able to do in the future what they have done so well in the past.

A few examples can be cited in which lower costs and less pollution have been achieved simultaneously. In 1971, five paper companies in the United States and Canada disclosed a new paper-making process which could save millions of dollars per year, reduce water pollution, and produce a higher quality product. The heart of the process was a new method for grinding wood into pulp, increasing the strength of the paper, and reducing waste and the amount of chemicals needed.[107] In other fields also, the amount of wastes per unit output have been improving: 20 years ago, canneries were producing 40 cases per ton of fresh peaches, whereas today they produce 55 cases; in 1928 cows produced 4.3 tons of manure in making 1 ton of milk, but by 1968 the ratio had decreased to 2.6 tons to 1 ton of milk; from 1943 to 1963 the wastes generated by the container board industry dropped from 0.45 ton per ton of final product to 0.21 ton.[108] In the energy-intensive aluminum industry, Alcoa in 1973 announced plans for a new plant which would reduce the energy required to produce aluminum by one third. Not only would this process have a favorable impact on resources and cut Alcoa's costs, but it would also eliminate undesirable fluoride emissions and create a better working environment for employees.[109] In 1973, the president of General Motors announced: "A great many

[107] *The Sacramento Bee*, April 15, 1971, p. F4.

[108] R. G. Ridker, "Population and Pollution in the United States," *Science*, June 9, 1972, p. 1089.

[109] *San Francisco Chronicle*, January 12, 1973.

of our plants will be on a closed system by 1980. In a majority of our assembly plants, water will not be returned to the sewer system; we'll just use make-up water. Probably when we're all through it will cost us less money and possibly we'll be cycling out useful by-products."[110]

These accomplishments are encouraging but should not lead us into a "technology-will-solve-all-our-problems" kind of optimism. Not only will we need major technological innovations in the future, we will also undoubtedy need innovations in our social institutions as well. In this total endeavor, engineers will play a prominent part, but so will economists, biologists, physicians, philosophers, and politicians. No doubt there will be certain technological exploits we could conceivably carry out but will choose not to. Through it all, we can be sure there will be much controversy, with the final result probably being that we will be better off than if the controversy had not taken place.

---

[110] P. Vanderwicken, "G. M.: The Price of Being 'Responsible'," *Fortune*, January 1972, p. 99ff.

TWO

Ethics and public responsibility

The nature of ethics

Ethics and morality are among the most difficult subjects to discuss. Opinions concerning these matters are frequently automatic, held on an *a priori* basis, and are never subject to verification until after the fact, and often not even then.

"What is justice?" asked Plato, and we are still seeking the answer today. The determination of justice in the social sense has come to mean, for many persons, the appropriate use of technology. To some critics, any use of technology leads to increasing alienation and dehumanization; therefore, technology is declared to be immoral. However, most people recognize the great improvement in the quality of human existence which has been produced through technology. For them, the issue is to maximize the gains and minimize the losses stemming from technology. Hence, they focus upon the principal practitioners of technology—the engineers—and wish to examine the moral nature of the decisions made by these practitioners.

Philosophers almost without number have attempted to dissect the subject of ethics in the hope that ethical behavior might be reduced to a science. Three principal doctrines predominate in ethical analysis: 1) the *objective* approach, in which it is believed that "goodness" or "rightness" are objective properties of human acts, just as mass or color are properties of physical objects; 2) the *subjective* approach, which requires reference only to an individual's inner feelings concerning the rightness of any action; 3) the *imperative* approach, which simply means that individuals use their personal ethical concepts as means for influencing the behavior of others.[1]

Each of the foregoing doctrines has its adherents, as do various combinations of them. Many persons believe in the objective theory,

---

[1] S. Toulmin, *The Place of Reason in Ethics* (Cambridge, England: The University Press, 1968), p. 5.

although they may not even be aware a theory exists which bears that name. They tend to believe that moral principles are self-evident and invariant. On the other hand, under the subjective theory, adherents usually feel they *know* what constitutes ethical behavior without reference to external rules. In an extreme form, the subjective approach turns up as the credo, "If it feels good—do it!" which has found favor with some. In a more reserved form, the popular "situation ethics" borrows from the subjective doctrine. As for the imperative approach, also known as "legalism," we all (or almost all) subscribe to it more or less, because systems of law and moral codes are expressions of this view. The noted philosopher Hans Reichenbach has put the imperative doctrine thus: "The fundamental ethical rules . . . are adhered to merely because human beings want these rules and want other persons to follow the same rules."[2]

Throughout history, there have been attacks upon legalism by those who believe laws and other codes to be inadequate guides to moral action. *Existentialism* and *situation ethics* are prominent examples of twentieth-century antilegalistic doctrines. In both of these, there is extreme emphasis upon the individual, and upon his right to determine his own course of action. However, a prominent feature of existentialism is also its emphasis upon alienation, whereas in situation ethics the emphasis is upon love (referred to by situationists as agape, or brotherly love).[3]

Situation ethics has aroused both great acclaim and great condemnation. During the 1960s, as civil disobedience over racism and war claimed more and more followers, situation ethics seemed to offer just the kind of moral authority which was needed to justify unlawful acts. Situationists assert that there is a higher moral imperative than the law, and that it is therefore sometimes necessary to break the law.

An example of such a case might be the necessity to take a gun away from a madman (commit a theft) to prevent him from killing others, or even to kill him (commit murder) if necessary to stop him. An example which was timely in the sixties was the use of the "sit-in" (commit trespass) to combat racial injustice. At a simpler level is the familiar instance in which it would be better to run through a red light if the alternative would be to sit still and be hit by a runaway truck.

Examples of the foregoing sort could be continued endlessly, and most people would probably agree that the circumstances of the examples were such that breaking the law was justified. But situationists go further. They assert there are no rules whatsoever—that there is only love as the guide to ethical action. Herein lies the source of the principal criticism of situation ethics, for who is wise enough to know

---

[2] H. Reichenbach, *The Rise of Scientific Philosophy* (Berkeley: University of California Press, 1951), p. 304.

[3] J. Fletcher, *Situation Ethics: The New Morality* (Philadelphia: The Westminster Press, 1966).

which course among alternatives is the most loving? Another criticism is that situationism may be used as a rationalization for selfish or evasive motives. The reply to this criticism by Joseph Fletcher, the principal spokesman for situation ethics, is, "Seek the best welfare and deepest happiness of the most people in the situation."[4] In this, situation ethics echoes the doctrine of utilitarianism, which in the eighteenth century declared that the object of ethical action is to produce the greatest good for the greatest number. However, even this laudable objective is criticized because it may cause damage to the welfare of some in order to promote the welfare of others. In the United States, much of the moral concern of the nation has been with how to avoid damage to the welfare of minorities who may otherwise lose out in an elective process which depends upon majority rule.

Most ethical systems seek to reduce the distress level in a society. In primitive societies, the notions of "duty" arise in response to the desire to avoid causing suffering or inconvenience to other members of the community. Ultimately, these duties are crystallized in taboos and are observed automatically. Thus are ethical codes of behavior born. And so it is with us today. Ethics constitute the basic codes of civilized behavior, without which our environment, as we know it, would be impossible. Such rules embody the basic constraints each of us agrees to practice in his relationships with others. We consent to these constraints in the knowledge that in so doing we make the existence of all, including ourselves, more agreeable. The basic role of ethical codes in society is well described by Stephen Toulmin, who says, "The concept of 'duty,' in short, is inextricable from the 'mechanics' of social life, and from the practices adopted by different communities in order to make living together in proximity tolerable or even possible."[5]

In much ethical discourse, certain ideas occur repeatedly. The situationist's desire to produce the greatest welfare for the most people echoes the utilitarian's principle which is 200 years older. The modern emphasis upon love is simply a restatement of the scriptural command to "love they neighbor." Immanuel Kant's Categorical Imperative to "Act according to a maxim which can, at the same time, be valid as a universal law,"[6] is hard to distinguish from the Golden Rule's requirement to "Do unto others as you would have them do unto you." These ideas represent the most fundamental guides to ethical behavior and represent the distilled wisdom of thousands of years of human experience.

The foregoing is not meant to imply that ethical codes or systems of law are infallible. All of us are aware of too many instances in which

---

[4] H. Cox (Ed.), *The Situation Ethics Debate* (Philadelphia: The Westminster Press, 1968), p. 254.

[5] *The Place of Reason in Ethics, op. cit.*, p. 136.

[6] I. Kant, *The Doctrine of Virtue* (Philadelphia: University of Pennsylvania Press, 1964), p. 24.

strict interpretation of the law results in injustice, or in which there seem to be no alternatives available to us which are free from undesirable consequences. A familiar example of the latter is the question of whom you would save from a burning building—your father or a medical genius who might benefit the world? Situation ethics would seem to suggest you should save the medical genius, since this would result in the greatest welfare for the most people. To complicate matters, it is just possible that you might be able to save both, but cannot be certain because no one can foretell the future. Furthermore, you must make your choice *now*—on the basis of inadequate information—and second-guessers of the future will be merciless in their attacks upon you if it seems to them you made the wrong choice. Of such stuff are moral dilemmas made. Ethical codes can be of help in most instances which confront us, but dilemmas do arise in which it seems there are no suitable alternatives. In such a case all external guides are helpless, and it is indeed necessary to resort to one's own internal values.

The greatest moral dilemmas in society are those related to war. Engineers, say some, are immoral if they work on weapons, because their clear moral duty is to oppose war by refusing to be involved in those activities which support it. The views of these critics are greatly strengthened today by the fact that the most recent war in U.S. history was an unpopular one. There have been "popular" wars in the past. World War II became instantly popular in the United States after the dramatic attack by the Japanese on Pearl Harbor. Prior to that time, the dominant U.S. attitude in the twentieth century was to stay out of wars. But after World War II started, there was great criticism of the United States, Great Britain, and France because they had not moved forcefully against Germany in the mid-1930s when that country first began making aggressive acts. If Germany had been stopped early, said these critics, World War II would never have happened, and millions of lives would have been saved. While the critics did not actually assert that the Allies' inaction was immoral, they did charge them with being incredibly stupid not to see what Hitler was up to.

Later attempts by the United States to apply the "lesson" of World War II led, of course, to the Cold War and ultimately to Viet Nam. As that tragic affair unfolded, more and more people came to see our involvement in Viet Nam as immoral, although it began with a reasonable amount of public support. The general belief at the beginning was that our readiness to oppose communist expansion was necessary in order to avoid World War III, as shown by the "lesson" of World War II. Antiwar groups later branded this view as a sorry coverup for naked imperialism. The U.S. public became more and more confused until it was finally paralyzed with its moral dilemma, especially as the underpinning assumptions of the Cold War began to appear obsolete.

The intention of the foregoing discussion is not to attempt to justify war but to create some sympathy for the perpetual human plight, which is how to select those courses of action that will produce the greatest human benefit. This is the activity which has dominated all others in the history of human development, and the most amazing thing about it is that it has been so successful. Over the centuries, freedom has increased, health has improved, the need for brutish labor has diminished, and options for human development have multiplied.

The good of society

It has been said that medicine has given mankind health, that the humanities have given mankind pleasure, and that technology has given mankind the time to enjoy both.

In many countries of the world today, men exist only in order that they may continue to exist. This was true in our own country not so long ago. There was little time for anything but work—12 to 14 hours of it each day, six days a week. It was not only our enlightened visions of right and wrong that caused the 40-hour week to become law in 1938. It was also a *necessity* as machines came, more and more, to replace men at routine tasks. If men had continued to work the 84-hour week of the nineteenth century, the result could have been only the concentration of jobs among fewer people, coupled with rising unemployment.

Ever since 1900, spendable income has continued to rise, so that it has been possible for most Americans to satisfy more of their wants while working fewer hours. Most of us have heard stories how our grandfather or great-grandfather would work all day cutting a wagonload of wood and then drive to town and sell it for a dollar. A dollar went farther in those days, but as Figure 2-1 shows, not *that* much farther. In 1909, workers in manufacturing enterprises earned an average of 19 cents per hour, with no fringe benefits; thus, the average weekly earnings were $9.74 for a 51-hour week.[7]

When the figures are corrected for the cost-of-living increase, it can be seen that such a worker existing in 1971 would have a purchasing power corresponding to a wage of 86 cents per hour ($44 weekly). Yet, the actual wage in 1971 was $3.57 per hour ($142.44 weekly), not counting fringe benefits. This increase in actual purchasing power is an improvement of 325 percent, during a period when the work week was being cut by 27 percent.

This immense improvement in only 50 years did not merely happen. Technology made it possible, and engineers are technology's principal implementers. Virtually every kind of innovation regarded by humans as beneficial has been subject to the influence of engineers: automobiles, airplanes, highways, television, and home appliances, to name some obvious ones. Some less obvious benefits include: newspapers,

---

[7] *The Economic Almanac: 1964* (New York: National Industrial Conference Board, 1964), pp. 54, 55, 63, 76, 77, 103.

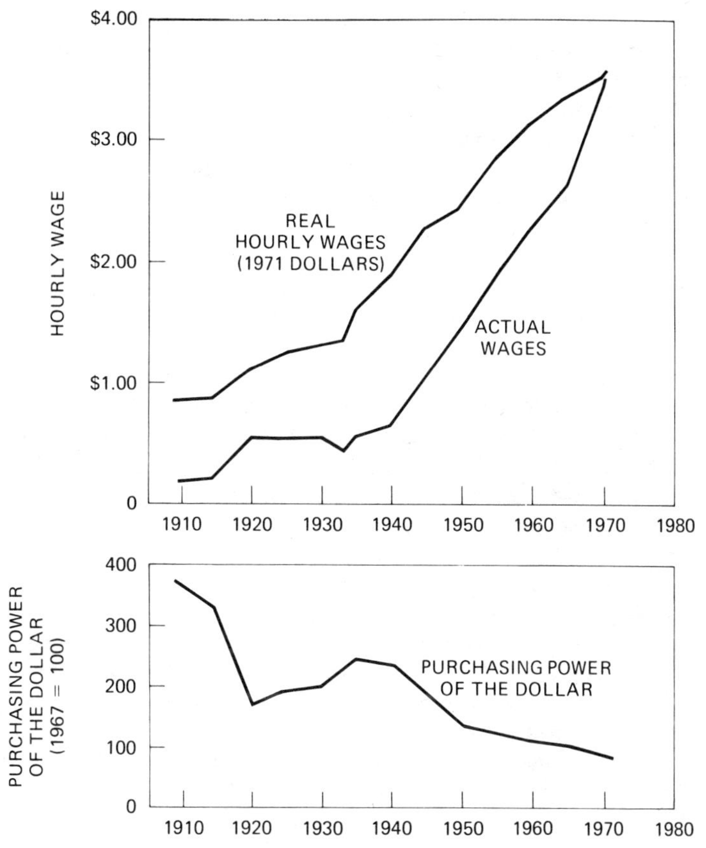

Fig. 2–1 Comparison between wages and cost of living from 1910 to 1971. Real wages represent actual purchasing power. Data refer to production workers in manufacturing. (Sources: *The Economic Almanac, 1964*. New York: National Industrial Conference Board, and *Statistical Abstract of the United States, 1972*. Washington, D.C.: U.S. Bureau of the Census.)

magazines, and books (printing machinery); food (harvesting, processing, and packaging machinery); new art forms (movies and television, or music for mass consumption, made possible by stereo and radio); and national security. Although not everyone agrees that engineering contributions to national security benefit mankind, most Americans believe that the production of nuclear missiles and other items for defense prevents their way of life from being disrupted by outsiders.

The freeing of man from a life of hard labor is a magnificent achievement. However, this release has not yet been accomplished throughout the world. In some Asian countries, three fourths of the population are needed just to produce enough food to support a society existing at the borderline of starvation. In the United States, less than 4 percent of the working population is required for the growing of food.[8] The options opened thereby for other, more enjoyable activities, are enormous. In a very real sense, the majority of the people in the United States can thank their scientists and engineers that they do not have to spend 14 hours a day laboring in a corn field.

These examples all underscore technology's impact upon productivity. The result of an increase in productivity is that more goods and services are available to more people with less human labor input than before. Thus, people are not only better fed and better housed than before and can afford better health care, but they also have a whole new range of options open to them for personal fulfillment. Rather than being forced into a life of bare subsistence, which was virtually the only course open to most people before the twentieth century, persons in industrialized nations have options available to them in education, a variety of careers, the arts, travel, sports, and entertainment beyond the dreams of those only a few generations ago.

The inverse relationship between productivity and poverty should be clear. As more goods and services are available to more people, the incidence of poverty decreases. The reduction in poverty in the United States in just the 10-year period from 1960 to 1970 is startling. In 1960, 22.2 percent of the U.S. population was below the "low-income level" of $3022; by 1970, only 12.6 percent was below the low-income level, which by that time had been raised to $3968 to correct for inflation.[9] In absolute numbers, there were 15.2 million fewer Americans below the low-income level in 1970 than in 1960.

Antitechnology

Countercultures of one sort or another have been with us throughout historical times, but within the last century they have taken on strong dimensions of antitechnology. This is not so much because of a hatred for technology itself, but because technology worships reason, whereas the counterculture enshrines emotion. Furthermore, technology supports the state, and counterculturalists generally believe that the state is obscene. Henry David Thoreau, who was a prominent counterculturalist in his time, declared that he was willing to accept the motto, "That government is best which governs least," but believed it should be altered to read, "That government is best which governs not at all."[10]

---

[8] *Statistical Abstract of the United States, 1972* (Washington, D.C.: U.S. Bureau of the Census, July 1972), p. 230.

[9] *Characteristics of the Low-Income Population, 1970* (Washington, D.C.: U.S. Bureau of the Census, Series P-60, No. 81, November 1971).

[10] C. Bode (Ed.), *Thoreau* (New York: The Viking Press, 1947), p. 109.

Counterculturalists reject dogma and organization, stress freedom of individual choice, and urge a return to simplicity. In their vision, "No coercion or delegation of responsibility occurs; the individual comes or goes, acts or declines to act, as he sees fit."[11] Above all, counterculturalists reject majority rule—the right of the majority to have its way over the minority.

The leading countercultural philosopher of today no doubt is Herbert Marcuse. His principal theme appears to be that technology has robbed man of his freedom because it supplies his wants so effectively that he ceases to seek any alternatives other than the one he now has.[12] Thus, to Marcuse and his followers, the technological state is a totalitarian state. Contrasted to this, there is a paradoxical vision in Marcuse's writings of a future perfect world in which technology has succeeded in transferring all unpleasant work from humans to machines. He says, for example, ". . . freedom indeed depends largely on technical progress, on the advancement of science." And again: "Is it still necessary to repeat that science and technology are the great vehicles of liberation, and that it is only their use and restriction in the repressive society which makes them into vehicles of domination?"[13] It is ironic that Marcuse, with this apparent faith in the beneficence of technical progress, should be one of the principal high priests of the counterculture, with its loathing for technology.

In the minds of two other social critics, Jacques Ellul and Theodore Roszak, no doubt exists as to the principal source of our troubles: it is technology. Ellul calls it *La Technique*, and Roszak calls it the *technocracy*.[14] By this they both mean the same thing, but it is far more than just technology. Ellul describes it as "the totality of methods rationally arrived at and having absolute efficiency . . . in every field of human activity."[15] Roszak says, "It is the ideal men usually have in mind when they speak of modernizing, updating, rationalizing, planning."[16] In their conviction that a culture guided by rationality has produced a meaningless, "plastic" existence, counterculturalists have rejected rationality as a guide, and have turned instead to instinct. They do not have

---

[11] G. Woodcock, *Anarchism: A History of Libertarian Ideas and Movements* (Cleveland: The World Publishing Co., 1962), p. 32.

[12] H. Marcuse, *One-Dimensional Man* (Boston: Beacon Press, 1964).

[13] H. Marcuse, *An Essay on Liberation* (Boston: Beacon Press, 1969), pp. 12, 19.

[14] The term "technocracy," as Roszak uses it, should not be confused with Technocracy, Inc., which aspired to control of society by "technical experts," and had a brief heyday in the 1930s. See *Understanding Technology*, by Charles Susskind (Baltimore: The Johns Hopkins University Press, 1973).

[15] J. Ellul, *The Technological Society* (New York: Vintage Books, 1967), p. xxv.

[16] T. Roszak, *The Making of a Counter Culture* (Garden City, N.Y.: Anchor Books, 1969), p. 5.

to think through the possible outcomes of moral issues in order to decide what to do; they just "know" what is moral and what is not.[17]

A prominent countercultural belief is that the American Indian knew how to live in harmony with nature, without technology. There is much to admire in Indian culture, but the belief that Indians rejected technology or that they possessed a greater ecological wisdom than modern society does not seem to be supported by the evidence. Indians of North America adopted the bow and arrow—a technological innovation—as soon as they discovered its superiority over the throwing stick. They adopted the steel knife as soon as it became available. Weaving and the mechanical loom were in use in America centuries before the white man arrived.[18] The vanished Hohokam built a complicated culture nearly 1000 years ago in Arizona based upon an irrigation system with hundreds of miles of canals. About the same time, the Anasazi of the Four Corners country built giant "apartment houses" having a thousand people in one structure, and constructed soaring towers reminiscent of medieval castles. But the greatest technological achievement of North American Indians was undoubtedly domestication of maize, or corn, which under cultivation increased spectacularly in size and productivity from its original wild state.[19]

Sadly, it appears that ancient Indian cultures have had their share of ecological calamities. For example, the theory has been advanced that the Hohokam of Arizona had to abandon their fields because their irrigation systems poisoned the ground with salt.[20] At Chaco Canyon, there is evidence that thick stands of pine and fir existed there when the people first arrived in 900 A.D. By 1100 A.D., they had felled the trees by the thousands to make roof beams for their huge pueblos. With the trees gone, the typical desert pattern of alternating wet and dry periods became much more devastating in its effects, contributing to the eventual abandonment of the region.[21] Finally, it is believed by some that the early extinction of certain species of animals in North America, including the mammoth, was caused by man's excessive hunting—a case of prehistoric overkill.[22]

Closely coupled with the foregoing view is the one that "nature knows

---

[17] C. A. Reich, *The Greening of America* (New York: Bantam Books, 1972), pp. 282–283.

[18] R. Underhill, *Pueblo Crafts* (Washington, D.C.: U.S. Bureau of Indian Affairs, 1944), p. 47.

[19] C. W. Ceram, *The First American* (New York: New American Library, 1972), pp. 219, 234, 327–334.

[20] D. E. Carr, *Death of the Sweet Waters* (New York: Berkley Publishing Corp., 1971), p. 131.

[21] R. Silverberg, "The End of the Anasazi," in *The American Indian*, R. F. Locke (Ed.) (Los Angeles: Mankind Publishing Co., 1970), p. 53ff.

[22] P. S. Martin, "Prehistoric Overkill," in *Man's Impact on Environment*, T. R. Detwyler (Ed.) (New York: McGraw-Hill, 1971), p. 612ff.

best," a theory which has been advanced by Barry Commoner as his Third Law of Ecology.[23] This notion has been disputed by another biologist, René Dubos, who reminds us that, "Hailstorms, droughts, hurricanes, earthquakes, and volcanic eruptions are common enough to make it obvious that the natural world is not the best possible world. . . ."[24] Dubos also reminds us that agricultural landscapes, regarded by most as beautiful and "natural," are actually man-created and man-managed. He offers the thesis—which is also the thesis of this book—that technology can indeed be tamed and humanized for the benefit of mankind.

Nature is the most important fact of life for mankind. We live in and are a part of nature. We depend upon it—utterly. We may forget this fact only at our peril, because if we sufficiently mistreat our Spaceship Earth we will have no escape from the consequences. But this is not the same as saying that man must not tamper with nature, that man is somehow excluded from nature's embrace. It is a mistake to assign a state of moral perfection to nature-without-man. For example, nature consists of a host of predator-prey relationships. These are usually viewed with approval by environmentalists as ecologically sound. But no one has yet thought to seek the view of the prey at the instant it becomes food for the predator. One wonders if its view of its personal environment at that fatal instant is a very satisfactory one. Alternatively, one wonders if the predator is convinced that its own environment is perfect after it has gone without food for a week during one of those frequent periods of scarce prey. These things can only appear "perfect" to the human—himself well fed and reasonably immune to becoming prey—who observes the system from the outside and finds it to his taste. We must face the fact that we want things certain ways because we find them pleasing and not because of their presumed state of natural perfection.

Responsibility

Since the mid-1960s, there has been an increasing amount of concern over the legal responsibility of manufacturers for product safety. The landmark event which is generally accepted as the beginning of this trend is the case of *Greenman v. Yuba Power Products, Inc.*, which established the principle of "strict product liability." Under this principle, a manufacturer is held liable for a product which proves to have a defect resulting in injury, even though there may be intervening

---

[23] B. Commoner, *The Closing Circle* (New York: Bantam Books, 1972), p. 37.

[24] R. J. Dubos, "Humanizing the Earth," *Science*, February 23, 1973, p. 769ff.

sellers.[25] During about the same period of time, there has also been a mounting debate over the moral obligations of engineers, not only with regard to safety, but also relating to other matters such as pollution, nuclear weapons, and war.

Undoubtedly the best-known critic of product safety (and of engineers) is Ralph Nader. It is his thesis that engineers identify their personal objectives too closely with those of the companies for which they work. He feels that the professional creed of engineers should require them to be independent of corporate directions.[26]

A more militant approach is that taken by the Committee for Social Responsibility in Engineering (CSRE). CSRE seeks to challenge the present orientation of engineering and declares that its objectives are to ". . . end unemployment and pollution and provide adequate medical care, housing, education, transportation and communication systems for all people." To implement these objectives, engineers are to be organized in groups, to oppose corporate power.[27]

These are difficult and important matters. The crux of the matter is: what is the engineer's role in a company? Is it proper for him to adapt himself to the corporation's directions? (Or to get out if he doesn't like them?) Or is it proper for him to assume a role of independence, reach a moral judgment on an issue, and then to demand that the corporation accept that judgment, while not relinquishing any claim to his salary? Presumably, if the latter course is chosen, the engineers in a company will have organized themselves into something like a union, withholding their services (that is, going on strike) if their wishes are not met. Implicit in this arrangement is the existence of a mechanism for reaching a collective moral judgment on an issue. However, to be honest, in reading the works of those who propose such measures, there appears to be an automatic expectation in the proposals that the groups, once formed, will adopt the moral views of the proposers. Such an expectation may be unrealistic.

Let us take the case of nuclear weapons. Many young people have made a major issue of the fact that they have lived all their lives under the shadow of the bomb and perceive this to be an unusual moral burden. The scientists and engineers who developed the bomb are condemned as immoral; otherwise, say the critics, they would have refused to work on such a devilish device. (Furthermore, they say, if the United States would just stop working on nuclear weapons, so would the Russians, the Chinese, the French, and so on.) During World War II, when the atomic bomb was developed, the United States believed itself to be in a desperate race with Germany to see which

---

[25] *Greenman v. Yuba Power Products, Inc.*, 59 Cal. 2d 453 (1963).

[26] R. Nader, *Unsafe at Any Speed*, rev. ed. (New York: Bantam Books, 1973), pp. 161–162.

[27] From CSRE statement of purpose, *Spark*, Spring 1973, pp. 1, 3.

would have the bomb first. Instead of condemning the United States because it was first, we should thank God it was so. We hardly need to ask what would have happened if Nazi Germany had been first, considering what transpired at extermination camps like Treblinka and Auschwitz. It is estimated that more than 1 million Jews were executed at Auschwitz alone. The total number of exterminations in Germany during World War II has been estimated as high as 7 million.[28]

To suppose that the United States could have safely declined to develop the bomb is only to indulge one's capacity for 20-20 hindsight, because we know now that Germany lost. In 1941, with France out of the way, Britain starving, Russia reeling on the ropes, and the United States on the sidelines, it appeared Germany had won the war. Except for the Japanese attack on Pearl Harbor, which brought the United States into the war, Germany may have won all of Europe. Many people think the United States would have been next.

Nuclear diplomacy on the world stage is beyond the scope of action of most engineers. Product safety is another matter. Virtually every engineer becomes involved with safety as a regular part of his professional activities. Few engineers would knowingly or uncaringly design unsafe devices. But safety frequently comes at increased cost, and the engineer must inevitably balance safety against cost.

If, to make a product safer, it must be made stronger and heavier, or must be equipped with extra devices, or made of superior materials, to the point that it is so much more expensive that few will buy it, what has been accomplished? The manufacturer may have to close its doors if it cannot sell its product, but it is hard to see how morality has been served thereby. A result of the company's closure may also be that some useful products have been denied to the public.

Sometimes a product can be made safer without increasing the cost; sometimes it cannot. It is suggested frequently that the extra costs of safety should come from corporate profits, but it is possible for a corporation eventually to go out of business if it fails to make a profit. Studebaker, Hudson, and RKO all have disappeared in the last two or three decades.[29]

Corporations are neither as perfect as portrayed by some, nor as evil as portrayed by others. They are simply the economic units in our society which are needed to supply the goods and services we need. No better means to serve this purpose has yet been found.

While corporations obviously would prefer to market safe products rather than unsafe ones, they feel justified in asking why they should

---

[28] W. L. Shirer, *The Rise and Fall of the Third Reich* (Greenwich, Conn.: Fawcett Publications, 1962), pp. 1259, 1267.

[29] For a highly readable account of the modern agonies of seven U.S. companies, see R. A. Smith, *Corporations in Crisis* (Garden City, N.Y.: Anchor Books, 1966).

voluntarily increase the safety of a product if the result is that sales suffer. The provision of safety belts in autos is a good example. In the early days of safety belts, most car buyers were either apathetic or antagonistic toward such devices. Even today, with greater acceptance of belts, many drivers refuse to use them. When the safety belt idea was new, it is easy to imagine what would have happened to a manufacturer who, concerned with his moral responsibility for safety, put belts into every car, whether wanted by the customer or not. There would have been a storm of customer protest. The only feasible method in such cases is for the government to enter as a third party, decide whether the public interest is vitally at stake, and if so, require all manufacturers to take equal action. Thus, the "body politic" has acted to bring about resolution of the moral question involved, without invoking the unrealistic expectation that one company should voluntarily act in an altruistic way, when the principal effect of the altruism may be only to give an advantage to the company's competitors.

The leveling effect of governmental action is indispensable in producing improvements in product safety, pollution reduction, and the like. It does little good to exhort an engineer to insist that his ideas on safety or pollution be adopted, if the effect would be to jeopardize his employer's welfare. If the engineer's action results in damaging his company, has he properly fulfilled the ethical obligations he assumed when he accepted employment? In return for a salary, there is an implied obligation that an employee will help advance the company's interests.

It is frequently not clear where the line is to be drawn between "safe" and "unsafe." Is a kitchen knife safe? Nearly 200,000 Americans are injured by kitchen knives every year. Is taking a bath safe? Over 100,000 are injured yearly in tubs and showers. Should one ride a bicycle? One million people are injured by bicycles each year in the United States.[30] Such hazards are commonly accepted by the public with little question. Each instance obviously carries great benefit, and the degree of risk is considered acceptable. The safety built into a product must be matched with the level of public acceptance. In cases which are sufficiently controversial, the acceptance level finally must be arbitrated by governmental action.

The individual engineer cannot be expected to assume a hard-line moral position which moves very far ahead of public opinion. If engineers as a group presumed to take on themselves the authority to act as moral judges for the rest of society and to provide or withhold certain items from the public, such action would be to assume totalitarian powers over others. The justification for applying the word "totalitarian" in this context lies in the fact that a small group—the

---

[30] *Final Report of the National Commission on Product Safety.* (Washington, D.C.: National Commission on Product Safety, June 1970), pp. 10–11.

engineers—would be making decisions for society which were not subject to review by elected officials or by the functioning of the market system. This arrogation of power would act to cancel the authority of the "body politic," which is supposed to decide questions of public policy. Furthermore, engineers possess no special qualities which make them superior moral authorities, to justify such a presumption to power.

Engineers have an obvious ethical obligation to the public when out-and-out violations of the law and/or safety codes occur or when known safety hazards are concealed. Even if moral obligations were to be set aside for a moment and the most selfish personal motives invoked, an individual engineer should ask whether his own long-range interests are well served by continued employment with a company which would knowingly conceal a safety hazard or violate the law. The appropriate course of action for the engineer is best phrased by the Board of Ethical Review of the National Society of Professional Engineers (NSPE):

> The engineer should make every effort within the company to have the corrective action taken. If these efforts are of no avail, and after advising the company of his intentions, he should notify the client [customer] and responsible authorities of the facts.[31]

After taking the obligatory action referred to above, the engineer had better be prepared to follow up either by fighting or by resigning. It should not necessarily be assumed, however, that one's career with the offending corporation is finished if a strong stand is taken. Sometimes the violation is the result of overzealousness on the part of lower echelons, and top management may be horrified to learn of the unethical and illegal acts that are being committed by their employees, presumably in the company interest. A good example is that of the Ford Motor Co., which is 1973 reported extensive violations—by its own employees—of the pollution control test procedures of the federal government. Evidently, in order to help test cars meet the emissions standards of the Environmental Protection Agency, Ford employees had given the cars unauthorized maintenance, such as replacement of spark plugs, cleaning carburetors, and adjustment of timing. When Ford's higher management discovered these violations, they reported the facts to the government and withdrew their application for certification.[32] Even these statesmanlike acts did not protect Ford from receiving a $7 million fine, however—all brought on by overzealous employees acting contrary to the public interest.

A distinction must be made between product safety and product quality. The former is the business of the public at large, whereas the

---

[31] *Opinions of the Board of Ethical Review* (Washington, D.C.: National Society of Professional Engineers, 1965), pp. 41–42.

[32] *San Francisco Chronicle*, February 14, 1973.

latter is between the company and its customers. In clear-cut cases involving safety, the engineer has an obligation to the public, but in matters involving product quality, the engineer only has the obligation to make the necessary facts known to management. As the NSPE's Board of Ethical Review has said: "If the public is misled as to the product's quality, the unfavorable reaction will be directed against the company."[33] Inevitably, cases will occur in which issues regarding quality and safety seem to be interrelated. The individual engineer must decide if the safety aspects involve violation of the law or serious concealment of hazard; if this is the case and if the company knowingly proceeds, no ethical alternative exists except to notify the customer and the authorities.

Many safety issues will involve differences of opinion, and some individuals may become exceedingly militant on seemingly irrelevant grounds. Even here, however, we must be careful, for something which seems fanciful today may not seem so tomorrow. An example is the issue of double insulation for electric power tools. Not many years ago, it would have seemed a foolish precaution to provide power tools with two layers of insulation, yet today the industry seems to be moving toward this requirement as standard. In the past, reliance in the United States was placed mostly on a third wire for grounding, but a 1969 study by Underwriters' Laboratories found that grounding was effective for only about 13 percent of the power tools in use. Many electrocutions have resulted from power tools in the United States, whereas in some European countries where double insulation has been broadly adopted, *no* electrical shocks have been reported for double-insulated tools.[34]

Fires in television sets have been a major problem. About 10,000 such fires occurred in the United States in 1969, mostly in color sets. About half of these fires affected only the sets themselves, but many escaped to cause financial loss, injury, and death. The reason for the high rate of fires in color sets is that they require extremely high voltages. The materials being used could not stand these conditions, and they broke down. Whether any moral dereliction on the part of engineers was involved, is unlikely. Even an extensive testing program might not have disclosed the problem. There were 20 million TV sets in the United States in 1969, and if 10,000 of them caught fire, the failure rate was about 0.05 percent. Thus, in a test program, it might be necessary to operate 2000 sets for a year before one catches fire. Thus, this particular problem may have occurred because it was beyond the state of anyone's knowledge that a danger existed. Once TV manufacturers were alerted to the problem, a crash program was adopted to upgrade

---

[33] *Opinions of the Board of Ethical Review, op. cit.*, p. 43.

[34] *Final Report of the National Commission on Product Safety, op. cit.*, pp. 26–27.

standards, even though costs would be increased. Regretably, according to the National Commission on Product Safety, a few manufacturers resisted the upgrading on the grounds that the hazard was "infinitesimal."[35]

Serious legal questions are involved in product liability. If faulty design of a product causes injury to someone, who is liable—the engineer or his employer? Court decisions seem to have held that the company offering the product is liable.[36] In one case, the faulty design of an aluminum lounge chair was held to be the cause of an injury that resulted in the loss of a finger; the manufacturer was held to be liable.[37] In another case, an infant was burned when a vaporizer near his crib caught fire. The vaporizer was not equipped with a cutoff and caught fire after the water boiled away. A $65,000 judgment against the manufacturer was affirmed, even though it had been shown that some other vaporizers on the market did not have cutoffs either.[38]

In much of the criticism directed at corporations, the technique of extreme advocacy is apparent. Ralph Nader's indictment of the auto industry, *Unsafe at Any Speed*, is an example. Another example is the literature of environmental organizations such as the Sierra Club. In this technique, one accentuates (and even exaggerates) the failings of the opponent, while neglecting to mention his virtues. Conversely, the virtues of one's own side are emphasized, and the failings omitted.

The criticism of the auto industry represents a good example of the foregoing. The critics cite failings of the auto manufacturers to introduce certain improvements voluntarily, principally those features relating to safety. They usually omit mention of the improvements that *have* been achieved over the years. For example, Ford first introduced padded dashboards and seat belts in 1956, long before Ralph Nader appeared on the scene. Autos also need far less maintenance than they used to. Only a few years ago, a lube and oil change was needed every 1000 miles; now it is every 6000 miles. Car finishes no longer need constant polish jobs. Seat covers used to be mandatory after a car was a few years old, but not any more. Tires are safer and last many times longer than they used to.

On the other hand, if a car manufacturer were to move too far ahead of public opinion, a reduction in sales could be his reward. Suppose one of the auto companies had voluntarily introduced pollution-control devices, say, in the early 1960s. Suppose the buyer had to pay a higher price, and in return would receive an auto which is harder to start, was more sluggish, and achieved poorer mileage (just like the autos of

---

[35] *Final Report of the National Commission on Product Safety, op. cit.*, pp. 13–14.

[36] D. W. Dodson, "Problems of Product Liability Claims," *Mechanical Engineering*, January 1965, p. 34ff.

[37] *Matthews v. Lawnlite Co.*, Fla., 88 So. 2d 299.

[38] *Lindroth v. Walgreen Co.*, 94 N.E. 2d 847 (1950).

the 1970s, in other words). It does not take much imagination to guess what the reaction of that buyer would have been. The public will accept such treatment only when it knows that every make of car is on an equal basis as to cost and performance.

The technique of advocacy is the method employed by lawyers in court trials, and it has spread through much of our social structure. This technique has obtained some spectacular gains during the turbulent sixties and seventies. Cars are certainly safer today than they were when *Unsafe at Any Speed* was first published. Much corporate and governmental behavior has been brought into question by critics such as Nader, the Sierra Club, and others, with improvements coming about as a result. Our method of government requires the forceful expression of public opinion to prevent us from sliding into retrogressive practices. On the other hand, extreme advocacy may have the effect of eroding public confidence in our social institutions. The incredibly tenacious manner in which the great depression of the thirties held on has been attributed to such a loss of confidence. The great challenge of our modern era is to keep both business and government constantly under searching public exposure and criticism, while simultaneously not damaging them beyond their ability to function.

The foregoing discussion has related mostly to the responsibilities of manufacturers, and thus is relevant particularly to electrical and mechanical engineers. However, civil engineers obviously have an enormous responsibility for public safety, because they design the structures upon which we all depend. Civil engineers have an enviable record of safety, but still, disasters have occurred.

One of the most sensational of such events occurred in 1928 when the St. Francis Dam in California burst without warning in the middle of the night and killed nearly 400 people.

Almost as soon as it had been completed and had begun to fill, leakage developed between the dam and the foundation. During its second year of service, several very large cracks formed in the dam. The water below the dam turned muddy, and William Mulholland, Chief Engineer of the Los Angeles Water Department, personally inspected the situation. He was relieved to find that the muddy water was caused by nearby road construction and decided the dam was not in immediate danger. That same night it collapsed. Mulholland, 73 years of age, publicly took full responsibility for the disaster and resigned his post shortly thereafter.[39]

Investigations showed that the dam site was intersected by at least one eathquake fault. Furthermore, when samples of the "bedrock" were tested by immersion in water, they were observed to change into a

---

[39] R. A. Nadeau, *The Water Seekers* (Garden City, N.Y.: Doubleday, 1950), p. 116.

mushy mass. In the urgency of Los Angeles' need for water, Mulholland had overlooked ordinary engineering precautions. An investigating committee concluded ". . . that the dam was constructed without a sufficiently thorough examination and understanding of the foundation materials upon which the dam was constructed." A coroner's jury found Mulholland responsible for an error in engineering judgment, but no charge of criminal negligence was made. The jury added that construction of a great dam "should never be left to the sole judgment of one man, no matter how eminent . . . ."[40]

A more recent incident of almost exactly the same type occurred in France in 1959; the Malpasset Dam collapsed and killed 421 people. In 1961, a government engineer was charged with involuntary homicide through negligence, and in 1964, his case came to trial. He was acquitted. The prosecution stated the engineer had been responsible for seeing that tests of the foundations were adequately carried out and charged that he had not carried them out correctly. However, one of the witnesses testified that the blame really should fall on the designer of the dam, who had been absolved of responsibility for the collapse.[41]

Hardly any bridge failure has received so much publicity as that of the Tacoma Narrows Bridge in 1940. The failure took place in broad daylight with cameramen present. Fortunately no lives were lost, as the bridge's violent undulations gave ample warning that something was about to happen. This bridge was the most slender large suspension bridge ever built.[42] It went down in a moderately strong wind, after having been in service for only four months. The cause of the failure was the creation of "Karman vortices," which were shed from the leeward side of the deck at a frequency coinciding with one of the natural frequencies of the bridge.[43] During and after the investigation, no charges of negligence were made, although it was noted that, "The builders of this bridge, being limited in funds, were anxious to build as inexpensive a bridge as possible in order to build any bridge at all." However, when the bridge was rebuilt, it was much less slender than before.[44]

---

[40] "Essential Facts Concerning the Failure of the St. Francis Dam," *Proc. Am. Soc. Civil Engrs.*, October 1929, pp. 2147–2163.

[41] *Engineering News-Record*, October 29, 1964, pp. 14–15, and December 3, 1964, p. 23.

[42] *Engineering News-Record*, November 14, 1940, p. 10. Width-to-span ratio for some major bridges: Verrazano Narrows, 1 to 41; Golden Gate, 1 to 47; Mackinac, 1 to 56; Tacoma Narrows, 1 to 72 (original) 1 to 47 (rebuilt). See *Engineering News-Record*, August 23, 1962, pp. 32–33.

[43] J. P. Den Hartog, *Mechanical Vibrations*, 4th ed. (New York: McGraw-Hill, 1956), p. 308.

[44] "Failure of the Tacoma Narrows Bridge," *Proc. Am. Soc. Civil Engrs.*, December 1943, p. 1568.

Some persons have insisted that engineers aim for no less than 100 percent safety, or that society should allow no new technological undertakings unless it can be guaranteed that there will be no undesirable side effects. Neither of these conditions is achievable, of course. Life cannot be made 100 percent safe, nor can any human being guarantee the future. We are always going to be faced with the presence of a certain amount of risk, although engineers, with their special technical knowledge of materials and the like, are in a better position to reduce risks than almost any other group. Furthermore, an engineer who is constantly thinking about matters such as safety and pollution is much more likely to produce designs that excel in these categories than is one who doesn't think about them.

Canons of ethics

Engineers have a well-developed set of canons. Some engineering societies have individually published their own. A set proposed by the Engineers' Council for Professional Development (ECPD) is reproduced in the Appendix.

Many of the duties prescribed by the Canons are also required by law, for instance, the one stating that an engineer will act ". . . as a faithful agent or trustee for each employer or client." In other cases they exceed what the law requires and may operate primarily to enhance the dignity of, and respect for, the engineering profession. However, no code of ethics should ever be allowed to become self-serving and to act only to enhance the profession. The objective of increasing public respect is justifiable only if it also increases public trust and confidence (which results in a public benefit) in engineers.

One obligation of a profession is the maintenance of high standards of conduct among its membership, with respect both to its members' relations with the public and to fair practices among fellow professionals. To this end, the engineering societies appoint special committees to review charges of unethical conduct against individual engineers. Usually, the action taken is merely a rebuke of the offenders and a caution against further offenses. In some instances, professional engineering societies have suspended or expelled wrongdoers. However, it appears that such sanctions are relatively rare, and some have doubted that they represent a very strong deterrent to malpractice.[45]

Many engineers were especially shocked to discover that engineers were involved in the payoff scandals in the State of Maryland, in 1973. To prevent the repetition of such events, a task force of the National

[45] W. G. Rothstein, "Engineers and the Functionalist Model of Professions," in *The Engineers and the Social System*, R. Perrucci and J. E. Gerstl (Eds.) (New York: Wiley, 1969), p. 91.

Society of Professional Engineers recommended that legislation be passed limiting all political contributions to $100, and that State Boards of Registration be empowered to enforce professional codes of conduct.[46]

The challenge

It is difficult to make very meaningful remarks about the future. One thing appears clear, however, and that is that engineers and scientists will become even more important to society in the future than they have been in the past. Major technical innovations will be necessary in order to provide energy, as our population continues to grow and our resources to diminish. Additional innovations will be necessary to avoid pollution and to recycle materials. Ways will have to be found to unclog and rebuild our cities. As our population growth begins to slow and finally to cease (as it must), technical innovation may turn out to offer a substitute for the economic stimulus which has been provided by simple population growth in the past.

The basic question to be answered is what kind of world we desire for the future. Until fairly recently, few have bothered to ask the question, although Aldous Huxley proposed an unsavory possibility in *Brave New World*, and George Orwell another one in *1984*. In the past, the prospect of increasing leisure and materialistic affluence appeared to be all that was necessary, in the minds of most. But the logical culmination of ever-increasing leisure is a life in which meaningful activity has disappeared, and all one's physical needs are supplied by robots. It is this prospect which so alarms Roszak. In *The Making of a Counter Culture*, Roszak claims he sees nothing at the end of the road we are presently following but ". . . Samuel Beckett's two sad tramps forever waiting under that wilted tree for their lives to begin." Roszak fears that the tree will not even be real, but plastic, and the "tramps" may turn out to be automatons.[47]

Roszak's vision, while apparently credible enough in the 1960s, appeared in the mid-1970s unlikely to be realized. Instead, it looked like society was going to have to do all the running of which it was capable just to stay in the same place. The rising expectations of numerous Third World nations, with their own legitimate claims on a share of the world's resources, seemed to guarantee that this would be so.

A list of the desirable qualities in the world we seek surely would include these:

1. Sufficient food to sustain life and health
2. Adequate health care for all

---

[46] "A Time for Action," *Professional Engineer*, December 1973, p. 18ff.
[47] *The Making of a Counter Culture, op. cit.*, p. xiv.

3. Freedom from danger, whether from pestilence, from accident, from other humans, from natural disaster, or from man's own technology
4. Freedom to select one's own destiny, subject to the limitation that another's like freedom not be infringed
5. Release of mankind from the need to perform routine labor
6. Availability of multiple opportunities for creative, intellectual, and recreational activity

In point of fact, we have not even managed to achieve the very first one of these—not for all people—and it is the most basic of all. On the other hand, we have made enormous advances on every one of them in the last century, mostly because of increasing use of technology. The trouble with Americans, says historian Daniel Boorstin, is that ". . . we have lost our sense of history. . . . We compare ours with a mythical Trouble-Free World, where all mankind was at peace . . . . We compare our smoggy air not with the odor of horsedung and the plague of flies and the smells of garbage and human excrement which filled cities of the past, but with the honeysuckle perfumes of some nonexistent City Beautiful." [48]

The gap between reality and the ideals we seek, though narrower than in past centuries, is great enough to provide challenge for all nations, including the United States. Says Peter Goldman, of *Newsweek*: "America's preeminent virtue, if her historians are right, may prove to be neither her wealth nor her grandeur, but her hardihood." [49]

---

[48] *Newsweek*, July 6, 1970, p. 28.
[49] *Ibid.*, p. 20.

THREE

Energy: the ultimate problem

In the early 1970s it became apparent that Americans could no longer take for granted that they were the richest people in the world. Sweden and West Germany had both passed the United States in the traditional measure of wealth—Gross National Product (GNP) per capita—and Japan was coming on strong.[1] This reversal occurred rapidly as the United States began to lose its competitive position in world markets, suffered its first negative trade balances in this century, and as a result was forced to accept dollar devaluation. The situation was seriously worsened as the United States became more dependent upon foreign sources of oil, causing enormous amounts of dollars to flow to oil-producing countries in the Middle East. Such foreign concentrations of dollars were believed by many to be the cause of pressures on the dollar which finally forced devaluation. Others, however, believed the dollar had been overvalued for years and the devaluation was long overdue.

As fuel oil and gasoline shortages began to appear in the United States, spokesmen in the oil industry announced that the only way to meet demand was to increase oil imports even more than before.[2] How the country was to pay for these imports, in the face of a trade balance which was already unfavorable, was not explained. It has been pointed out that the United States has successfully retained a competitive posture in world markets in only two areas: "high-technology" products and agricultural products. In virtually every other field—motor vehicles,

---

[1] *Sweden NOW*, No. 4/1973, p. 8. In 1973, Sweden's GNP per capita was 108 percent that of the United States', West Germany's was 102 percent. Japan's ratio to the United States was only 15 percent in 1960, but by 1973 Japan had passed Great Britain (51 percent), had reached 62 percent of the U.S. GNP and was still rising.

[2] *The National Energy Problem: The Short-Term Supply Prospect* (Houston, Texas: Shell Oil Co., June 1, 1973).

steel, textiles, raw materials, and "nontechnology-intensive" manufactured products—the United States has lost ground.[3]

In the past, the United States achieved its position as the world's wealthiest country through a combination of fortuitous circumstances. The country was blessed with nearly every kind of resource that could be wanted: abundant water, metals, and energy; vast forests; and some of the most favored agricultural lands in the world. Coupled with these, it had a political system which encouraged innovation and had periodic influxes of immigrants who were willing to fill the laboring ranks and make the system go. As the twentieth century moved into its final quarter, it looked as if the United States had lost nearly all these advantages. Russia possessed vastly greater quantities of water, forests, metals, and coal than did the United States, the mid-East had the bulk of the oil, the United States clearly had no monopoly on innovativeness—even her "high-technology" products were being challenged—and the Japanese seemed to be working harder than anyone else in the world. Only America's vast agricultural productivity seemed still to be preeminent.

One of the most fundamental resources among those listed above is energy. Without energy, we cannot produce the goods we want, cannot provide adequate health care, cannot heat or light our buildings, cannot transport our goods or ourselves, and cannot even harvest our crops. The United States possesses abundant energy reserves, as will be seen, but they are of kinds which come at a special cost, or which cannot be used without extensive additional research and development. The preferred energy sources are in short supply. Even though these preferred sources may not actually run out for decades, we cannot produce energy from them at a fast enough pace to keep up with the demand. We cannot even convert over to other sources quickly enough to avoid unpleasant strains.

Technology has faced enormous challenges in this century and has had startling successes. The achievement of flight—first into the air and then into space—is clearly one of them. Instant communication around the world is another. The freeing of mankind from brutish labor is probably the finest of all to date. Now, technology faces its greatest challenge: to move from a depletable energy base, that is, from dependency on fossil fuels, to an energy base which is perpetual. Upon this achievement, all else depends.

The U.S. energy budget

As can be seen from Figure 3-1, nearly 80 percent of energy in the United States in 1971 was supplied by burning natural gas or oil. This

---

[3] F. Schulman, "Technology, the Energy Crisis, and Our Standard of Living," *Mechanical Engineering*, September 1973, p. 16ff.

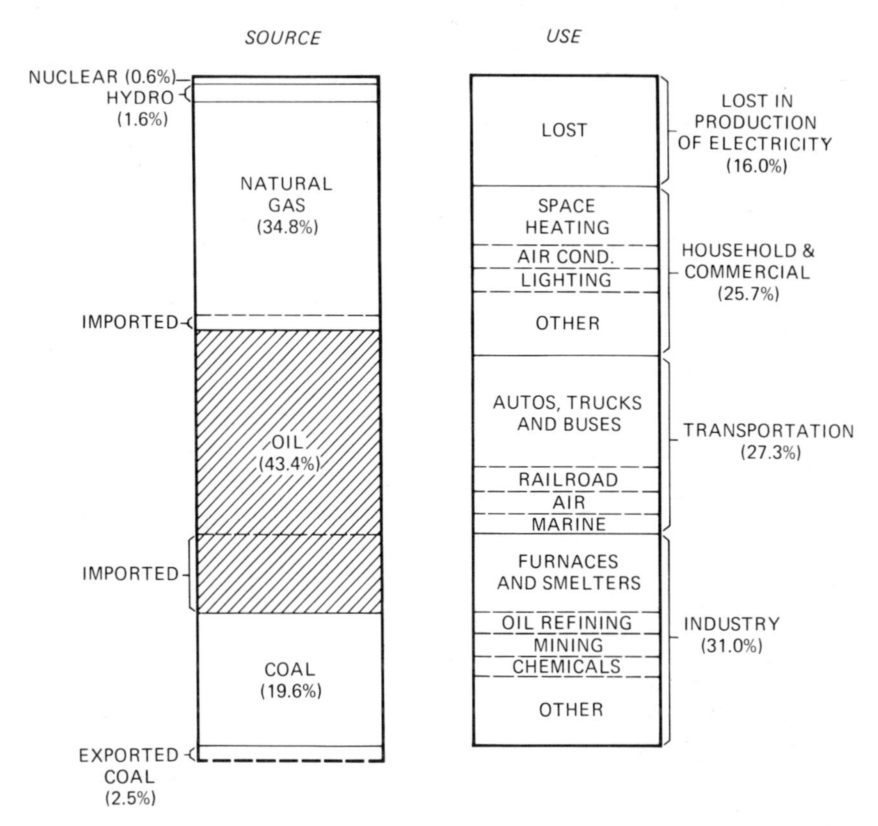

Fig. 3–1 Energy budget for the United States in 1971. Total energy usage: 63.2×10^{15} Btu. (Source: Earl Cook, Texas A&M University. See *Science*, December 8, 1972, p. 1081, and *Scientific American*, September 1971, pp. 138–139.)

situation cannot be changed rapidly, partly because of the capital investment required for conversion to a different fuel and partly because the only other fuel available in sufficient quantity is coal, which has been a bad polluter. The amount of energy available from hydroelectric installations is minor and will remain minor because most of the attractive hydro sites have already been developed. The fraction from nuclear energy was almost invisible in 1971 but was expected to increase as more nuclear plants come on line in the 1970s and 1980s.

Even if a sudden new supply of crude oil had become available when fuel and gasoline shortages first appeared, matters would have improved only slowly, because the United States did not have the refining capacity to handle a major increase. During a period of three or four years in the early 1970s, the refining capacity of the United States remained practically static because of governmental regulations, opposition by environmentalists, and corporate uncertainty regarding the kinds of fuels the refineries should be designed to produce.[4] Hence, it appeared the country would have to increase its importation of products already refined abroad, but this merely raised new questions regarding foreign refinery capacity, availability of shipping, and adequacy of U.S. ports to handle the load.

It takes three to four years to build a refinery, and five to seven to build a nuclear power plant. If we expect to rely more on coal, our most plentiful fossil fuel, an even longer delay is in store because new methods for sulfur removal will have to be proved out first, before a significant number of new plants can be designed and built. Simply to switch back to conventional coal burners means that we are back to polluting the atmosphere with sulfur. Most of the improvement in SO_2 pollution has been achieved by switching from high sulfur coals to other fuels such as low-sulfur oil and natural gas. Even though potential means were known for removing SO_2 from the stacks of coal-burning plants, these were still in the early stages of development when the energy shortages arrived. The sulfur could be removed if the coal were to be converted to gaseous or liquid form, but gasification and liquefaction plants also were only in the developmental stage when the shortages came.

On the "use" side of the budget, it comes as a surprise to most people that so much of our energy goes into heating, transportation, and industry. In the minds of most, electricity dominates the energy picture, yet only about 25 percent of our energy is used in electrical form. Furthermore, a large amount of energy is lost in generating electrical energy, if the electricity is used ultimately only to produce heat. The overall efficiency of an electrical power plant is about 35 percent, as compared to 70 percent when fuel is converted directly to heat.[5] If electricity is used for purposes other than heating, however, it has no effective competitors, as when it is used to produce light, or to run a computer.

The "transportation" part of our energy budget is only about 15 percent efficient, partly because internal combustion engines are not especially efficient in the first place and partly because they are required to operate across wide ranges of load.

Another kind of energy budget is that of the entire earth, shown in Figure 3-2. The principal message conveyed by this figure is that the energy received by the earth from the sun is enormously large—almost 30,000 times greater than man's entire use of energy. Other energy flows to the earth's surface are tiny in comparison. If man can ever succeed in tapping even a small part of the solar energy flowing to the earth, his energy problems will be solved. Unfortunately, the sun's energy is

---

[4] *Los Angeles Times*, January 25, 1973, p. 26. (Full-page ad by Mobil Oil Co.)

[5] *Science*, December 8, 1972, p. 1079ff.

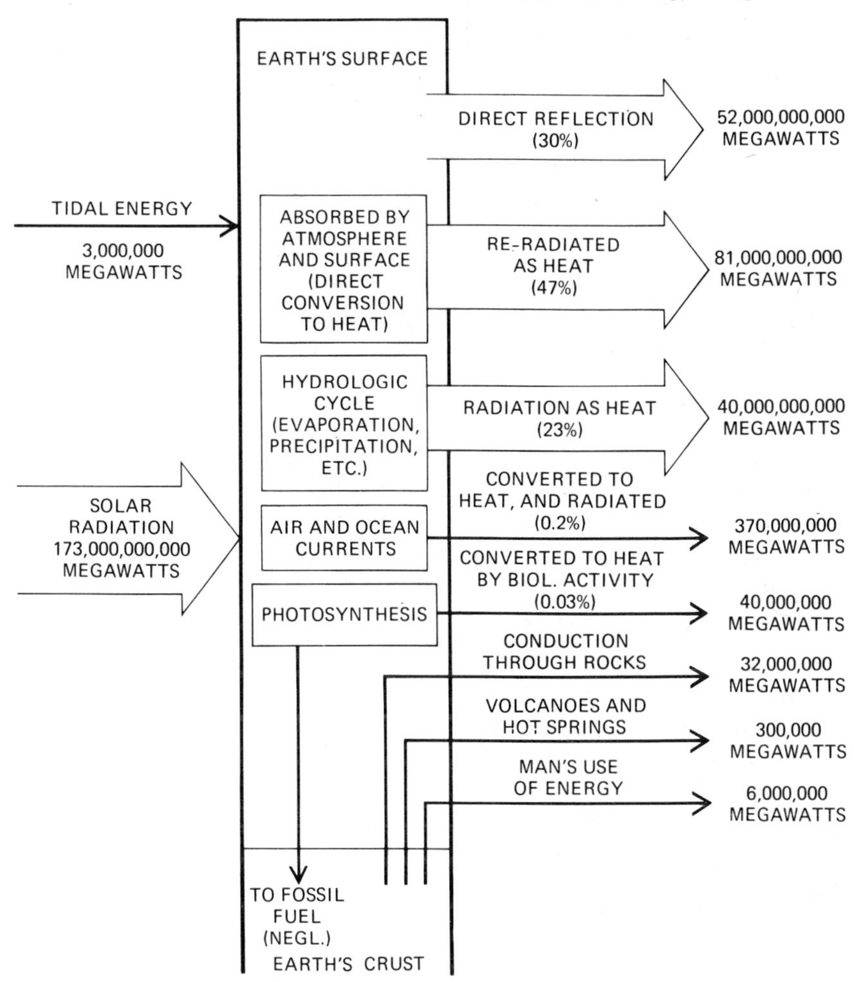

Fig. 3–2 Energy flow of the earth. (Source: M. King Hubbert, "The Energy Resources of the Earth," *Scientific American*, September 1971, pp. 62–63).

very dilute and is intermittent. Nevertheless, solar energy is a promising source.

Concern has occasionally been expressed that man's conversion of fossil fuels to heat might cause a noticeable warming of the earth, perhaps enough to alter its climate. As can be seen from Figure 3-2, man's production of energy is too tiny to produce such a result, although climates might be altered locally, as in large cities. If man succeeds in utilizing solar energy, an obvious benefit is that no net change in the energy balance of the earth takes place; man merely collects the energy and concentrates it where it is needed. (Local climatic alterations could still be produced, however.) On the other hand, if a major energy source such as controlled fusion is perfected, man's activities could eventually result in a significant additional heat load for the earth.

The fossil fuels

Oil, natural gas, and coal are our principal sources of energy, and they almost certainly will remain so through the rest of this century. Many attempts have been made to predict how long these resources will last. If one assumes that demand will continue to grow exponentially, as it has ever since World War II, then our reserves of everything except coal will disappear in a very few years. But, no matter what the "demand" might seem to be, we can never use more than the actual amount available in the supply, and the supply clearly will not continue to grow as fast as in the past. We will continue to make new oil and gas discoveries, of course, and increases in the price of energy will provide a further stimulus to exploration. However, the success rate in the discovery of new oil wells in the United States has declined in the years since World War II. The drilling of 64 unprofitable wells was required for each significant discovery in 1963, instead of the average of only 26 wells per discovery which prevailed in 1945.[6]

If one adds to the amount of known reserves an estimate of the amount of future discoveries, and also estimates the future demand, then it is possible to establish a probable date by which all reserves, discovered and undiscovered, will be used up. One such estimate is as follows:

	Energy equivalent of initial supply ($kWh \times 10^{12}$)	*Year by which resource is 90% depleted*
Crude oil (U.S. less Alaska)	275	2000
Crude oil (world)	2240	2020
Natural gas (U.S. less Alaska)	407	2015
Coal (U.S.)	5920	2400

Source: J. Holdren and P. Herrera, *Energy, A Crisis in Power* (San Francisco: Sierra Club, 1971), p. 145. (Source: M. K. Hubbert, *op. cit.*)

Even if Alaskan oil is added to the foregoing, the picture does not change much. The recoverable oil in the Alaskan fields is estimated at

[6] *Scientific American,* September 1971, *op. cit.*, pp. 64–65.

20 billion barrels. In the early 1970s, the United States was using about 5 billion barrels of oil per year; thus, if we had to rely solely on oil from Alaska, it would last only four years, even without allowing for increased demand.[7] As pointed out in an earlier chapter, the United States can expect 100 million more people by 2015, so it is foolish to suppose there will not be increased demand.

Much of the potential for oil and gas discoveries lies in offshore areas, both on the continental shelves of the world, and on the continental "rises" which represent the transition zones from the abyssal plains to the continental shelves. Scientists at the Woods Hole Oceanographic Institution believe the rises may contain even more oil and gas than the shelves, but the rises have not been explored by drilling because of the deep water in which they lie.[8]

As of 1970, the United States was using twice as much natural gas per year (22 trillion cu. ft.) as it was discovering (11 trillion cu. ft.), if we leave out Alaskan discoveries (26 trillion cu. ft.).[9] The natural gas industry claimed the shortfall in new discoveries primarily stemmed from unnaturally low prices which were set by the Federal Power Commission. If gas prices had been higher, the industry said, then demand for gas would not have risen so high so fast. Also, gas companies would have been encouraged more to look for new deposits.[10] In the meantime, plans were going forward for large-scale importation of liquefied natural gas (LNG) from abroad. In just one project, an investment of $1.7 billion was committed to the construction of a pipeline and liquefaction plant in Algeria ($738 million), unloading and regasification facilities in the United States ($277 million), and the construction of nine LNG ships ($700 million). A boom in shipbuilding was touched off by the need for LNG tankers, each of which is a collection of giant cryogenic tanks to keep the gas at $-260°F$ while it is carried across the ocean. Ironically, the French were the first to construct such tankers, even though the technique of liquefying natural gas was developed by the United States during the space program.[11]

In the case of oil, too, construction of giant tankers and port facilities are central matters. Oil can be transported in a "supertanker" at less than half the cost of transportation in a tanker of conventional size. Supertankers may be more than 1000 feet long and have a draft of 80 feet. Consequently, they need special deepwater terminals, frequently

---

[7] L. Rocks and R. P. Runyon, *The Energy Crisis* (New York: Crown Publishers, 1972), p. 22.

[8] K. O. Emery, "Oil on the Shelf," *Oceanus*, Spring 1973, p. 11ff.

[9] A. Austin, B. Rubin, and G. Werth, *Energy: Uses, Sources, Issues* (Livermore, Calif.: Lawrence Livermore Laboratory, May 30, 1972), pp. 3–17.

[10] E. Faltermayer, "The Energy 'Joyride' Is Over," *Fortune*, September 1972, p. 99ff.

[11] "The Hottest Item on the Shelf," *Fortune*, April 1973, p. 60.

consisting of mooring buoys and platforms with pumping and pipeline facilities located several miles offshore in deep water. It has been claimed that supertankers operating into such terminals would have less environmental impact than conventional tankers using existing ports. With large tankers, fewer of them will be operating, which lessens the chance of collision. Also, if tankers tie up at offshore terminals they can stay out of congested shipping lanes, which again reduces the chance of accident.[12]

In view of the difficulties with natural gas and oil, the United States has turned to its most plentiful energy source: coal. There are three main problems: 1) to burn coal directly is to invite back our past pollution problems with SO_2; 2) most of the known coal reserves in the United States lie in beds more than 1000 feet deep and are costly to mine; 3) the readily recoverable coal is mined by surface stripping, which constitutes a special kind of environmental pollution.

A method for avoiding sulfur pollution is to turn the coal into gaseous form before it is used. The sulfur is removed during the gasification process, and the gas either is consumed on site for power generation or is pumped into the natural gas pipeline delivery system. However, many successive steps are required to produce methane (CH_4), which has a high heating value and is the most highly desired gas. Basically, the process involves bringing coal into contact with steam at pressures as high as 70 atm and temperatures to 1500°C. Production of low heating value gas from coal in one step is possible, but the heat content is only a small fraction of that of methane, and it cannot economically be pumped into natural gas pipeline systems. (Natural gas is mostly methane.)

Several methods for coal gasification have been tested on a laboratory scale, and an additional method has been used commercially in Europe. It has been proposed that as many as 36 coal gasification plants could be in operation in the United States by 1985, with a total production capacity of 3 trillion cu. ft. per year. Each plant could cost $350 million for a total investment of $77 billion.[13] A problem is sure to develop in constructing gasification plants on such a massive scale. The problem is to obtain a sufficient water supply, because water is the source of the hydrogen which combines with the coal to make CH_4.

Instead of gasifying coal, it has been proposed that coal should be converted to methanol (CH_3OH) instead. The coal is first treated with steam to produce CO and H_2, as in gasification. The CO and H_2 are then combined, under heat and pressure, to make CH_3OH. Historically, meth-

---

[12] "Deepwater Ports: Issue Mixes Supertankers, Land Policy," *Science*, August 31, 1973, p. 825ff.

[13] Gasification: A Rediscovered Source of Clean Fuel," *Science*, October 6, 1972, pp. 44–45. Also, "Power Gas and Combined Cycles: Clean Power from Fossil Fuels," *Science*, January 5, 1973, p. 54ff.

anol has been widely used as a motor fuel, is less polluting than gasoline, and can even improve the performance of gasoline when mixed with it to about 15 or 20 percent.[14]

In the final analysis, whether coal is gasified, liquefied, or burned directly must depend upon the relative economics of the alternate processes. Just to burn coal directly will cost more than in the past because of the need for sulfur removal. A former director of the U.S. Bureau of Mines has estimated that a cost of $4 per ton of coal would be required for stack-gas removal of sulfur. However, he estimated it would cost $9 per ton to ship low sulfur coal to the industrial midwest from western sources. He also pointed out that conversion of coal to gas or oil is costly in an energy conservation sense, because 25 percent of the energy in the coal is used up in the conversion process.[15]

Along with the renewed interest in coal, there emerged simultaneously a strong resistance to strip mining from environmental groups such as the Sierra Club. The U.S. Department of the Interior estimated that, in 1965, 3.2 million acres of the United States had been surface mined by one method or another. The principal surface-mining methods are: 1) open-pit mining, as exemplified by giant copper mines and gravel pits; 2) area strip mining, conducted on relatively flat land by scraping off the "overburden" to get at the desired commodity underneath; 3) contour strip mining, practiced in mountainous terrain by cutting directly into an exposed vein on a hillside, leaving a wide bench behind which resembles a road cut; 4) auger mining, in which a narrow bench is made, wide enough to accommodate a giant augering machine, which drills holes deep into the mountainside, boring out the coal. The last three methods are commonly used for obtaining coal, and account for 40 percent of all acres disturbed by strip mining. Sand, stone, and gravel pits account for the bulk of the remainder.

Two thirds of the area in the United States strip mined for coal—nearly 1 million acres—were located in four states in 1965: Kentucky, Ohio, Pennsylvania, and West Virginia. Of these acres, the U.S. Department of the Interior estimated about 40 percent needed no further reclamation. They had already been reclaimed, either by nature or by deliberate reclamation projects. But for the remainder—nearly 600,000 acres—reclamation was judged to be needed.[16]

In Europe, strip mining is extensively used, and reclamation is included as a matter of routine. The topsoil is stripped off first and saved. After mining, the cut sections are backfilled, the topsoil is

---

[14] T. B. Reed and R. M. Lerner, "Methanol: A Versatile Fuel for Immediate Use," *Science*, December 28, 1973, pp. 1299–1304.

[15] E. F. Osborn, "Coal and the Present Energy Situation," *Science*, February 8, 1974, pp. 477–481.

[16] *Surface Mining and Our Environment.* (Washington, D.C.: U.S. Dept. of the Interior, 1967), pp. 33–34, 110–111.

restored, and the area is planted to hay and grain. The job is made easier because the land is relatively flat. Nevertheless, restoration was judged to cost $3000 to $4500 per acre, which might increase the price of a ton of coal by 10 or 20 percent.[17] However, estimates for reclamation by West Virginia authorities are as low as $260 per acre for contour mining, and only $52 per acre for auger mining.[18] In the opinion of some experts, steep slopes should not be strip mined at all. In one survey of strip mining in Appalachia, no successful reclamation was reported on slopes greater than 28 degrees.[19]

A novel possibility exists for turning coal to gas which does away with the need for mining. This is to gasify the coal in place, referred to as *in situ*. The process is especially of interest for deep coal veins, because mining such beds is hazardous, expensive, and results in only about 25 percent recovery. Holes are drilled into the beds, and a high explosive charge set off in each hole to create fracture zones. Steam is pumped to the beds or, alternatively, the beds are set on fire underground, and water is piped into them to make steam. A low-heating-value gas is created, collected at the surface, and probably used on-site for electric power generation. The advantages are that less surface disturbance is caused, and dangerous deep mining is avoided.[20] A disadvantage is that slumping might occur in regions where large quantities of coal have been turned into gas below the surface. Slumping of the surface in mining regions is a frequent problem.

Other fossil fuel sources are those in oil shale and tar sands. It has been estimated that the U.S. oil-shale deposits—mostly in Colorado and Utah—contain 6000 billion barrels of oil, but only a small fraction of this is considered economically recoverable. A major problem with obtaining petroleum from oil shale is to handle the tremendous amount of waste produced.

Oil shale processing was still under development in the 1970s. On the other hand, production of oil from tar sands was in the commercial phase. Canada is estimated to have 400 billion barrels of oil in tar sands. In 1973, a facility in Alberta was producing 50,000 barrels of high quality crude oil per day from this source.[21]

---

[17] J. F. Stacks, *Stripping* (San Francisco: Sierra Club, 1972), pp. 95–98.

[18] S. M. Brock, *Auger Mining for Coal in Southern West Virginia—Costs and Benefits* (Morgantown, W. Va.: West Virginia University, March 1972), p. 6.

[19] D. B. Brooks, "Strip Mining in East Kentucky," in *Appalachia in the Sixties*, D. S. Walls and J. B. Stephenson (Eds.) (Lexington, Ky.: University Press of Kentucky, 1972), pp. 119–129.

[20] E. F. Osborn, *Science, op. cit.*

[21] S. D. Bechtel, Jr., *Energy for the Future: New Directions and Challenges.* Speech delivered at 54th Annual Convention of the American Gas Association, October 16, 1972, Cincinnati, Ohio.

Nuclear energy

During the 1960s, the electric utility industry made massive plans for construction of nuclear power plants, requiring an investment of billions of dollars. But by 1972, only a quarter of the plants projected for completion in that year were actually in service. Delays had developed for a variety of reasons, some technical, some environmental. The biggest issues had to do with safety.

The matter of routine radioactive releases from nuclear plants had for a time been a major controversy. It ceased to be an issue when the Atomic Energy Commission (AEC) tightened its requirements to the point that the individual dose at a plant boundary could be no greater than 5 millirems (mrem) per year, whether from gaseous or liquid effluents. Since the natural background radiation to which we are all exposed is 100 to 200 mrem per person per year, most knowledgeable observers felt that this aspect of the nuclear safety problem could be considered under control. Nevertheless, there were still some who felt that the dose at a nuclear plant boundary should be zero.[22]

A far more serious safety issue is involved with the potential results of a major reactor accident. If some accident were to occur which interrupted the flow of coolant to the reactor—rupture of a pipe, for example—the reactor would shut down immediately, but residual heat generation would be sufficient to melt the reactor unless an emergency core cooling system (ECCS) were brought into action. Nuclear reactors are equipped with such systems, but the unanswerable question is: what happens if the ECCS fails to work? Nuclear plants are also equipped with giant reinforced concrete shells designed to confine steam and radioactive fission products at high pressure in the event of an accident. But it is not possible to prove that the containment shell, let alone the ECCS, will indeed work as planned without fail.[23] Even if the containment shell works, there is little certainty concerning the ultimate fate of a melting reactor core as it settles downward through its foundation.

The danger of a loss-of-coolant accident is not that the reactor would turn into an atomic bomb. The danger comes from the possibility of a "conventional" explosion, which would cause radioactive materials to escape and be airborne over a large region. Even if a major radioactive release were to occur, there is argument over the consequences to human health. Some critics have claimed that thousands of deaths

---

[22] A. M. Weinberg, "Social Institutions and Nuclear Energy," *Science*, July 7, 1972, p. 27ff.

[23] "Nuclear Safety: AEC Report Makes the Best of It," *Science*, January 26, 1973, pp. 360–363.

would result within a few weeks. Other experts assert that the results would not be so extreme, although they admit that a likely result would be to raise the incidence of cancer in the region.[24]

Research programs have been directed at attempting to answer nuclear energy's nagging safety questions, but at the time of writing this book these had not produced usable results. Until more information could become available, about the best that panels of experts on nuclear safety could recommend was the nuclear plants be located away from urban areas.[25] One scheme for remote siting is to locate nuclear plants in offshore areas. Plans for the world's first floating nuclear power plant, to be located off the Jersey coast, were developed in the mid-1970s. Such a plant could either produce electrical power directly, or could be designed to produce hydrogen from water. The hydrogen could then be sent ashore in pipelines or in tankers and used directly as an energy medium.[26]

Other major safety issues in nuclear power have to do with transporting spent fuel elements back to chemical plants for reprocessing, and the problem of final disposal of radioactive wastes. If nuclear energy expands as envisioned, there will be thousands of shipments of spent fuel by railroad in the United States per year. The fuel elements will be hot, both in the thermal and radioactive senses. They will have to be enclosed in special casks equipped to dissipate heat during transit and will have to be shielded. The casks intended for such use have been designed to withstand almost any accident, but if something should occur which ruptures a cask, local radioactive contamination would occur. One proposal to minimize this particular hazard is to retain the fuel elements at the plant site for a long period of time so that they cool more before shipping. Another proposal is to locate nuclear plants and reprocessing plants adjacent to each other so that no shipment is necessary.

As for ultimate nuclear waste disposal, there are many proposals but no fully agreed upon solutions. The AEC has proposed disposing of the wastes in deep salt mines, on the theory that the presence of the salt shows that the beds have not be in contact with water for millions of years. Another proposal is to pump the wastes in liquid form into an explosively formed cavity in deep silicate rock layers. When the capacity of the cavity has been reached, the wastes would be allowed to boil dry, melting the surrounding rock, and sealing the wastes in

---

[24] T. Alexander, "The Big Blowup Over Nuclear Blowdowns," *Fortune*, May 1973, p. 317ff.

[25] *Nuclear Power Safety in California* (Sacramento: Assembly Science and Technology Advisory Council, May 1973), p. 3.

[26] L. Lessing, "The Coming Hydrogen Economy," *Fortune*, November 1972, p. 138ff.

place.[27] Waste disposal is one of the thorniest problems in nuclear power because some of the wastes will remain dangerous for thousands of years. A proposal has been advanced which would reduce the need for such immensely long storage. Since the long-life components of the wastes are materials such as uranium, neptunium, and plutonium, the proposal is to remove 99.9 percent of these elements from the wastes by chemical separation and "burn" them in a reactor. The wastes remaining would then have a troublesome life of 700 years or so, instead of thousands.[28] Even so, 700 years would seem like a long time to maintain vigilance over atomic wastes, in view of the fact that the United States has been a nation for only 200 years.

Another controversy in nuclear energy concerns the long-term availability of uranium. Uranium occurs naturally in two principal isotopes—^{235}U and ^{238}U. ^{235}U is a "fissile" isotype, meaning that it will fission if struck by a slow neutron, releasing energy, various fission fragments, and more neutrons. ^{235}U is the main fuel of today's reactors but constitutes only 0.7 percent of naturally occurring uranium. ^{238}U which makes up the other 99 percent of natural uranium, is a "fertile" isotope, meaning that it can be converted into a fissile material upon being struck by neutrons. The fissile material produced in this case is plutonium-239, which in turn can be used as a reactor fuel. The conversion of ^{238}U to ^{239}Pu goes on all the time in nuclear reactors. If less ^{239}Pu is produced than ^{235}U consumed, the reactor is called a *converter*. If more ^{239}Pu is produced than ^{235}U consumed, the reactor is called a *breeder*. The advantage of the breeder reactor is that it makes ^{238}U available as a fuel. Hence, there is great interest in developing practical breeder reactors.

In a converter, only about 1 percent of the uranium is turned into energy; in a breeder, it is estimated that 50 percent or more can be turned into energy. If only converters are placed into commercial use, it has been claimed that our uranium reserves would last only 35 years or so, if nuclear reactors were used to supply all our electricity at the 1970 level. Breeder reactors could extend the supply for 1500 years or longer. However, in the estimates just offered, there is an implicit assumption that the supply of uranium is that which is available at $5 to $10 per pound. If the cost were allowed to go as high as $500 per pound, by exploiting low-grade ore, the uranium supply has an estimated life of 10,000 years, even if only converters are used. A fuel price of $500 per pound might make it appear that electric power would

[27] J. J. Cohen, A. E. Lewis, and R. L. Braun, "*In Situ* Incorporation of Nuclear Waste in Deep Molten Silicate Rock," *Nuclear Technology*, April 1972, p. 76ff.

[28] A. S. Kubo and D. J. Rose, "Disposal of Nuclear Wastes," *Science*, December 21, 1973, pp. 1205–1211.

cost 100 times as much as if the price were $5, but such is not the case. It has been estimated that power costs would only double under the foregoing fuel cost increase, since plant investment costs and operating expenses represent a large share of power generation costs.[29] A complicating factor is that extensive mining of low-grade ore might come at unacceptable environmental cost.

Development of breeder reactors in the United States has lagged behind similar work in Europe. The United States was the first country to produce power from a small experimental breeder in 1951, but it then ran into trouble with an early attempt to construct a larger experimental breeder, the Enrico Fermi plant in Michigan, and the operation never became successful. In 1973, the Russians, British, and French all announced that developmental breeder reactors were beginning operation in their countries, while the United States was several years away from completing its first full-sized commercial breeder.[30]

The advantage of the breeder is its fuel economy. But critics have been skeptical of the assumptions made in the economic analyses made of the breeder. Furthermore, they say, the extremely high operating temperatures proposed for breeder reactors make them even more subject to accident than today's reactors. Also, with breeders, the large amount of plutonium produced would have to be guarded closely against theft, because terrorists conceivably could steal some to make a crude atomic bomb.[31] However, the same danger exists with conventional reactors because they also produce a certain amount of plutonium.

Even though nuclear energy has been controversial, this particular source of energy is on its way to providing a substantial share of U.S. electricity in the last part of the twentieth century. About 25 plants were operating in 1972, with 117 more planned or under construction. In spite of the troublesome safety questions, a properly operating nuclear plant produces far less environmental pollution than a fossil-fueled plant. Nuclear plants also cause less mining disturbance because a smaller volume of fuel is required. About the only form of environmental pollution emanating from a properly operating nuclear plant is waste heat rejected in the cooling water. Even this form of pollution is not necessary if additional capital investment is put into giant cooling towers. Cooling towers eject the waste heat directly into the atmosphere, instead of into lakes, rivers, or oceans, and thus avoid alteration of the biological balance of surface waters.[32] Thus, as with

---

[29] *Energy, A Crisis in Power*, *op. cit.*, pp. 60–61.

[30] "U.S. Fast Breeder May Become More Problem Than Solution," *Engineering News-Record*, March 8, 1973, p. 20ff.

[31] "The Fast Breeder Reactor: Signs of a Critical Reaction," *Science*, April 28, 1972, pp. 391–393.

[32] "Fission: The Pro's and Con's of Nuclear Power," *Science*, October 13, 1972, pp. 147–149.

most things in life, nuclear power comes as a mixed blessing and poses major unresolved dilemmas for society.

Geothermal energy

Seven countries in the world have power plants which utilize natural steam and/or hot water from the earth to generate electricity.[33] In 1973 the largest such plant in the world was in northern California, with a capacity of 396 MW. A plant near Lardarello, Italy, produced 380 MW. The California site was planned for expansion to 900 MW by 1980—about the size of a typical large nuclear power plant.[34]

Geothermal resources exist in three forms: steam, hot water, and hot rock. Existing plants have used the first two of these, and methods have been proposed for utilizing the third. In a plant which uses steam, the steam drives a turbine which is connected to a generator in the usual manner. The turbine must be designed for very different operating conditions than in a conventional power plant, however, for the working temperatures and pressures are lower than in the usual fossil-fuel plant. But sources of dry steam are rare, and most geothermal plants will have to utilize hot water mixed with steam. In these, the steam is either separated from the water before it is used or the heat from the steam/water mixture is transferred to a secondary working fluid such as isobutane. Heat transfer to a secondary fluid may be attractive anyway, since many hot water sources are heavily mineralized and cannot be readily discharged to the surface. Thus, the mineralized water may have to be reinjected into the earth after passage through a heat exchanger. Without controls such as reinjection some geothermal plants could release as much sulfur into the air as fossil-fuel plants.

The potential of heat recoverability from hot rock is considered to be very much larger than that from hot water. Hot rock lies within a few thousand feet of the surface in many places in the world, especially in the western United States. One proposal for utilizing this heat is to drill parallel holes deep into the hot rock and to create a large zone of cracks at the bottom of the holes by hydraulic pressure, a technique developed by the petroleum industry known as "hydrofracturing." Water would be pumped down one hole and become heated by circulating through the fractured zone. If things work as planned, the water would rise through the companion hole and be utilized at the surface in the same manner as hot water from natural sources.[35]

---

[33] The countries are Iceland, Italy, Japan, Mexico, New Zealand, Russia, and the United States. See "Geothermal Energy: An Emerging Major Resource," *Science*, September 15, 1972, pp. 978–980.

[34] *Pacific Gas and Electric Progress*, October 1973, p. 4.

[35] "Dry Geothermal Wells: Promising Experimental Results," *Science*, October 5, 1973, pp. 43–44.

Estimates vary concerning the amount of energy ultimately available from geothermal sources. Some experts make very conservative estimates, predicting that each thermal site may last only 30 or 40 years under intensive use. Using geothermal energy is really not much different from using fossil fuels, they assert, because each site is essentially a thermal deposit containing a limited amount of heat. Using it, then, is to be regarded merely as another form of mining. Other experts are more optimistic and claim that the recoverable energy in the hot rock of the western United States may be on the order of 10^8 MW-centuries. Since the total installed electrical capacity of the United States in the mid-1970s was about 400,000 MW, then if the optimists are right, we could meet all of the U.S. electricity needs for centuries from geothermal energy alone.

The U.S. Department of the Interior has been conducing explorations in the Imperial Valley region of southern California, which has extensive hot water sources. Estimates made by Professor Robert Rex of the University of California, Riverside, are that this source alone may have a magnitude of 30,000 MW. The total hot water potential of the United States, excluding hot rock sources, has been estimated to lie between 10^6 and 10^7 MW—several times larger than our total installed electrical capacity of the mid-1970s.[36]

Solar energy

Nearly all of the energy we use comes (or came) from the sun. Our fossil fuels are stored forms of solar energy carried over from millions of years ago. Hydroelectric power is made possible because the sun's energy conveniently transports water from the oceans to the mountaintops for us to use. Only those tiny portions of our energy coming from geothermal and nuclear sources are not basically from the sun.

Interest in using solar energy has fluctuated over the years. A few houses utilizing solar energy have been built, but an incentive for greater use has been lacking because fossil fuels have been so cheap. A resurgence of interest in solar energy has recently occurred because conventional forms of energy will no longer be cheap.

Some of the promising ways in which solar energy might be produced are 1) in small units for utilizing the sun's energy in heat form to produce heating or cooling for houses and small commercial buildings; 2) in small power plants to convert the sun's heat to electric form, as by generating low pressure steam to operate turbines; 3) in units for direct conversion of solar energy to electricity by photovoltaic cells;

---

[36] A. W. Weinberg, "Long-Range Approaches to Resolving the Energy Crisis," *Mechanical Engineering*, June 1973, pp. 14–18.

4) in large electric generation plants, either through use of turbine technology or by using photo cells.

The reason for making a distinction between small and large systems is that many of the technological details of the two would differ markedly. Small thermal systems would utilize low temperatures, whereas large ones would need high temperatures and thus would require lens systems or parabolic reflectors to concentrate the sun's energy. Small systems would utilize roof areas for energy collection and so would not require that vast new regions of land be locked up in huge solar collection arrays. On the other hand, large thermal plants would be more efficient than small ones because their operating temperatures would be higher.

A major problem in solar systems is to find a way to store energy until it is wanted. Batteries could be used to store electrical energy but are not economically attractive. Various methods for storing heat energy directly have been used or proposed, including storage in hot water tanks, hot rocks, and molten salt mixtures. One novel proposal for energy storage is to use the sun's energy to make hydrogen from water, either by electrolysis or chemically, and then store the hydrogen. When the energy is to be recovered, the hydrogen could be burned or used to operate a fuel cell.[37]

Hydrogen not only is attractive for storage purposes but may offer possibilities as a delivery medium, since it can be sent through pipelines or shipped by truck or tank car. If the hydrogen were used in fuel cells, an interesting by-product would be fresh water, which could be used instead of thrown away. Hydrogen could be burned directly for space heating and also could be used to run internal combustion engines.[38] In the latter case, obviously no HC, CO, or CO_2 would be produced, but NO_x would have to be controlled because nitrogen from the air would still be involved.

The eventual future use of hydrogen is full of uncertainties, and many challenges exist for engineering development. Currently, hydrogen is far more expensive than natural gas, but as natural gas supplies diminish, hydrogen will appear more attractive. A basic problem is how to generate the enormous volumes of hydrogen that would be required to duplicate today's usage of natural gas. If this amount of hydrogen were to be generated by electrolysis of water, it is estimated that an electrical capacity of 1 million MW would be required, which is nearly three times larger than the total U.S. electrical capacity in 1973. However, there are ways other than electrolysis to obtain hydrogen from water. Chemical reaction sequences are under development for production of hydrogen which use heat directly so that no intermediate pro-

---

[37] Weinberg, *Mechanical Engineering, op. cit.*, pp. 14–18.

[38] J. O'M. Bockris, "A Hydrogen Economy," *Science*, June 23, 1972, p. 1323.

duction of electricity is necessary.[39] In any event, the huge amounts of energy needed to make hydrogen would have to come from a basic source such as nuclear energy, geothermal energy, solar energy, or controlled fusion. However, if a biological method for production of hydrogen could be developed, such huge generating facilities might not be necessary. Methods for direct production of hydrogen from water by algae, stimulated by sunlight, have been investigated. Research results have only been preliminary, however, and are far from practical utility.[40]

Many people fear that hydrogen is an unusually dangerous fuel. It is dangerous, of course, as are natural gas and gasoline. However, all three of these substances are handled routinely in pipelines and tank car shipments—not without occasional accident. A special danger of hydrogen is that it requires an extremely low level of energy to ignite it, which makes it especially susceptible to spark hazard. On the positive side, its combustion does not produce CO, a substance which has been responsible for many deaths.

Of other ways to utilize solar energy, the method which is closest to practical utility is to use it to heat and cool homes. Air conditioning is an especially attractive application, since the time of greatest need is also the time of greatest solar energy availability. A refrigeration system employing absorption is one method which is undergoing development. A reversible cycle, which in summer would "pump" heat from a cool region (the house's interior) to a warmer one (the outdoors) could be used in winter to pump heat in the other direction, except that now the cool region would be the outdoors and the warm one the house's interior. When the sun's energy is not available, back-up systems employing fossil fuels (or hydrogen) would be necessary. Since a large amount of our total energy goes to space heating or cooling, such solar-driven systems would have a "stretching" effect upon our fossil fuels.

One major advantage of small units located at the points of use is that no additional areas of the earth's surface are locked up in solar collectors than that already used by homes and other buildings. However, if we go to large-scale solar plants, the land requirements would be enormous: 40 square miles for a 1000 MW power station. (It has been pointed out, however, that 40 square miles is also the amount of land that would be strip mined for coal over a 30-year period to supply a coal-fired plant of equal size.[41] In a large solar plant, lenses or mirrors would be required to focus the sunlight in order to achieve high enough

---

[39] D. P. Gregory, "The Hydrogen Economy," *Scientific American*, January 1973, pp. 13–21.

[40] "Hydrogen: Synthetic Fuel of the Future," *Science*, November 24, 1972, pp. 849–852.

[41] *San Francisco Chronicle*, August 23, 1973, p. 8.

temperatures to reach the desired efficiency. The sun's energy would be concentrated onto coated pipes containing a heat-conducting medium. The pipe coatings would be highly absorbent to the solar wavelengths but would have low emissivity in the infrared range so the incident heat would be mostly retained. Such coatings have already been developed for space applications and may ultimately constitute another "spin-off" from the space program for widespread human use.[42]

An intriguing possibility for direct conversion of sunlight to electricity is to use photo cells. Silicon cells have been used to power space systems but are very expensive. A possible alternative to silicon is cadmium sulfide, which would probably be adaptable to mass production but would be less efficient than silicon (6 percent, compared to 13 percent) and would have a short life. Gallium arsenide cells appear to possess high efficiency (18 percent) but are even more expensive than silicon. A very large number of cells would be needed to generate large amounts of electricity; the mind boggles at the thought of 40 square miles of Arizona desert covered with photocells. Nevertheless, the possibilities in this direction have appeared attractive enough to stimulate pilot research projects.[43]

Another unusual scheme for tapping solar energy is to utilize temperature differentials in the sea. Most of the sun's energy falls on the oceans, warming the surface layers; the net effect is that the ocean can be regarded as a giant heat storage unit. In tropical regions the surface temperature is generally about 25°C while at depths of 3000 feet or so the temperature may be only 5°C. These conditions are kept nearly constant by the sun's energy, which causes warm surface currents to travel toward the poles, while cold water from the polar regions moves slowly through the depths toward the equator. In a power plant, the warm surface water would heat a working fluid such as ammonia, causing it to boil. The vapor would operate a turbine, and the working fluid would be condensed by cooling it with cold water drawn from the depths. With such a small temperature differential, the achievable efficiency might be only 2 or 3 percent. Nevertheless, the potential economy of this type of plant was sufficiently attractive (the "fuel" is free) to support several research projects in the United States in 1973.[44] One team of researchers has estimated that the energy contained in the Gulf Stream could generate 75 times the output of all U.S. electric power plants.[45]

---

[42] "Solar Energy: The Largest Resource," *Science*, September 22, 1972, pp. 1088–1090.

[43] "Photovoltaic Cells: Direct Conversion of Solar Energy," *Science*, November 17, 1972, pp. 732–733.

[44] "Ocean Temperature Gradients: Solar Power from the Sea," *Science*, June 22, 1973, pp. 1266–1267.

[45] D. F. Othmer and O. A. Roels, "Power, Fresh Water, and Food from Cold, Deep Sea Water," *Science*, October 12, 1973, pp. 121–125.

The idea of using temperature gradients in the ocean to produce power is not exactly new. D'Arsonval first suggested it in 1881. A French engineer, Georges Claude, actually constructed a working power plant based on ocean temperature gradients in 1929 in Cuba. Claude used the sea water itself as the working fluid. Even though the temperature of the water was far below water's normal boiling point, it was made to flash into vapor by lowering the pressure. Claude's plant was an economic failure, but changes in technology and in fuel economics have caused his ideas—as well as interest in solar power generally—to be revived.[46]

Two other minor forms of renewable energy exist: wind power and tidal power. A few systems for utilizing these resources are already in existence, and further development is likely. However, the amount of power available in these categories is small compared to the sun's energy.

Controlled fusion

If man can successfully produce usable energy from controlled fusion, the full promise of energy from the atom will have been realized. Fusion offers the prospect of unlimited energy resources with little or no pollution and without most of the safety worries of today's nuclear reactors. The trouble is that the feasibility of controlled fusion has not yet been established, despite decades of effort, mostly by the United States and Russia. Even after the process has been experimentally demonstrated, it will probably take another two or three decades before fusion plants can start delivering commercial power because of immensely difficult engineering problems.

Fusion operates at the opposite end of the atomic scale from fission. Fission uses the heavy elements like uranium, thorium, and plutonium. Fusion uses the light elements like hydrogen and lithium. When fusion occurs, energy is released along with neutrons. It is the energy process of the sun as well as of the hydrogen bomb.

Two isotopes of hydrogen occur commonly in nature, while a third, called tritium, exists in nature only in trace quantities. The most common hydrogen isotope is our normal hydrogen with a mass number of 1. The other common isotope is deuterium, or "heavy hydrogen" (symbol: D, instead of H) with a mass number of 2. Tritium (symbol: T) has a mass number of 3. Deuterium is the principal intended "fuel" for controlled fusion. Even though the ratio of D to H in nature is only 1:6500, the total amount of deuterium in the world's oceans is so great that it could supply our energy needs for millions of years. Furthermore,

---

[46] G. Claude, "Power from the Tropical Seas," *Mechanical Engineering*, December 1930, pp. 1039–1044.

economical methods for obtaining deuterium from water are already known.[47]

There are two potential fusion reactions of interest: one between two deuterium nuclei (the "D-D" reaction) and the other between nuclei of deuterium and tritium (the "D-T" reaction). For fusion to take place, a minimum temperature of 400 million degrees K is required for the D-D reaction and 50 million degrees K for the D-T reaction. Since the D-T reaction requires a lower temperature, it is the reaction likely to be achieved first. However, it requires tritium as a fuel, and it appears that tritium will have to be produced from lithium by neutron bombardment. Once the D-T reaction is initiated, the neutrons produced by the reaction would strike a surrounding lithium blanket and generate more tritium automatically. It has been estimated that the earth's supply of usable lithium is only about equal to the supply of fossil fuels.[48] Hence, the ultimate achievement of the D-D reaction is of fundamental interest.

At the very high temperature of fusion, the atoms inside the reactor will all have been stripped of their electrons and will be present as a cloud of charged particles, called a *plasma*. Confinement of this plasma by magnetic fields has been the principal focus of fusion research work, since normal materials obviously cannot contain temperatures of millions of degrees. One of the engineering problems of fusion reactors will be to sustain supercold temperatures in the cryogenic coils of the magnets which generate the magnetic fields, only a few feet distant from the core of the reactor which will be millions of degrees hotter.

Fusion reactors should be safer than fission reactors because if something occurs which interferes with the conditions for sustaining the process, the process simply stops. Pollution will be almost nil since the product of the fusion reaction is nonradioactive helium. Unfortunately, tritium is radioactive but is a much less dangerous substance than the products of fission plants. Nevertheless, it will have to be controlled, because tritium combines with oxygen just as hydrogen does to make water and can pass into the food chain.[49]

An idea for fusion which does not require magnetic confinement is to deliver an enormous amount of laser energy to a frozen D-T pellet so rapidly that the necessary fusion conditions of temperature and density are achieved by causing the pellet to implode. Methods for delivering pellets in a rapid stream would have to be developed, as well as appropriate means for extracting the energy produced. However, the major barrier to laser fusion at the time of writing this book was the

---

[47] S. Glasstone, *Sourcebook on Atomic Energy*, 3rd ed. (New York: Van Nostrand Reinhold, 1967), pp. 178–182, 195–196, 360, 540–543.

[48] M. K. Hubbert, "The Energy Resources of the Earth," *Scientific American*, September 1971, pp. 61–70.

[49] "Magnetic Containment Fusion: What Are the Prospects?" *Science*, October 20, 1972, pp. 291–293.

unavailability of a laser big enough to do the job. The largest laser in existence was able to deliver only 600 joules, whereas a 10,000-joule laser would be necessary, capable of delivering its energy in less than a nanosecond.[50]

Considering all the problems to be solved, it is clear that controlled fusion cannot come into use in time to benefit the present generation or perhaps even the next one. Millions, probably billions, of dollars (or rubles, or whatever) will be spent by this generation, not for its own benefit, but for the benefit of future generations. If controlled fusion is indeed achieved, our energy problems can be considered solved. Such a magnificent enterprise should be regarded as evidence of humanity's concern for persons yet to be born.

Energy conservation

One of the greatest challenges to engineers is to bring about greater efficiencies in the use of energy. This activity lies in their natural realm and is one of the things they do best. Greater efficiencies in energy use could have been achieved in the past, but sufficient incentive to do so was lacking because energy has been so cheap. Even so, in some sections of industry notable improvements have been achieved. For example, from 1900 to 1973, the efficiencies of electrical generation plants have been increased from 5 percent to nearly 40 percent. The maximum theoretical efficiency obtainable from a thermal cycle is governed by the Carnot equation: Efficiency $= (T_1 - T_2)/T_1$, where T_1 is the absolute temperature of the working fluid at the input to the cycle, and T_2 is the absolute temperature at the outlet. In a modern steam turbine, T_1 might be 810°K (1000°F) and T_2 about 310°K (100°F), giving a theoretical Carnot efficiency of about 62 percent. However, the Carnot cycle is impractical, and a compromise cycle must be used. Furthermore, a boiler cannot convert *all* of the latent heat of its fuel to usable energy; some goes up the flue. A heat exchanger can never be 100 percent efficient, nor can turbines or generators. Thus, the overall efficiency of a system is always considerably less than the theoretical level. One of the major ways in which the efficiency of a thermal cycle can be increased is to raise the inlet temperature T_1, and much of the effort of engineers in recent times has been expended in this direction.

Another promising method for increasing energy conversion efficiency is to eliminate the use of a thermal cycle altogether and to go to direct energy conversion. One such method is to use a magnetohydrodynamic (MHD) generator. In MHD, a hot, partially ionized gas from a com-

---

[50] Weinberg, *Mechanical Engineering, op. cit.*, p. 15.

bustion chamber is made to pass through a magnetic field, causing an electric current in the gas which is collected at electrodes. The overall efficiencies of MHD plants have been predicted to go as high as 50 to 60 percent because of the absence of intermediate conversion steps. As of 1973, MHD had not as yet been developed to the practical stage, but ambitious development efforts were underway in many countries including the United States, Russia, and Japan.[51]

Other impressive increases in efficiency have been achieved in the past, and these help to indicate useful directions for the future. A nearly fourfold increase in rail transportation efficiency occurred in the 1950s because of replacement of steam locomotives by diesel engines. Since about 1940, there has been a large-scale shift from inefficient incandescent lighting to more efficient forms such as fluorescent and mercury vapor lamps. About 70 percent of the lighting in the United States is now provided by fluorescent lamps. Because of greater efficiencies, the nation's effective lighting was tripled between 1960 and 1970, with only a doubling in energy required.[52]

As can be seen from Figure 3-1, the industrial use of energy is the biggest portion of the overall energy budget. Economies have been systematically achieved in this sector in the past, and more will undoubtedly be achieved in the future. For example, the energy required to produce a ton of steel was reduced by 13 percent between 1960 and 1968, primarily because of new blast furnace designs.[53] In an earlier chapter, mention was made of a new process for making aluminum which reduces the energy requirement by one third. In some industries, managers have been startled to discover that simple things like steam leaks were wasting much energy and causing considerable financial loss as well. In one case, a proposed new million dollar boiler was found to be unnecessary after energy use was tightened up.[54]

Transportation is our second biggest energy user, and within this category autos, trucks, and buses account for the biggest share. In recent decades there has been a pronounced shift in the direction of less efficient forms of transportation because of greater convenience or greater speed. Autos have grown bigger and more powerful, with a consequent decline in miles per gallon of fuel. Trucks have displaced railroads in the movement of much of the nation's freight. Airplanes have taken over from railroads in the movement of passengers and

---

[51] A. L. Hammond, W. D. Metz, and T. H. Maugh, *Energy and the Future* (Washington, D.C.: American Association for the Advancement of Science, 1973), pp. 25–28.

[52] C. M. Summers, "The Conversion of Energy," *Scientific American*, September 1971, p. 149ff.

[53] *Energy and the Future, op. cit.*, p. 135.

[54] D. H. Dawson, "Why Not Save Energy Costs by $2 Billion a Year?" *DuPont Context*, No. 2/1973, pp. 17–19.

have even been assuming an increasing share of freight shipments. Comparative efficiencies for different transportation modes are given in the following tables:

Freight transport	*Btu per ton-mile*
Pipeline	450
Railroad	670
Truck	2800
Airplane	42000

Passenger transport	*Btu per passenger-mile*	
	Intercity	*Urban*
Bus	1600	3800
Railroad	2900	
Automobile	3400	8100
Airplane	8400	

Source: E. Hirst and J. C. Moyers, "Efficiency of Energy Use in the United States," *Science*, March 30, 1973, pp. 1299–1304.

Obviously, shifting of freight to railroads could achieve a significant saving in energy, although modifications in price structures and government regulations would probably be needed to accomplish such a change. For intercity travel, airplanes use more than twice as much energy per passenger-mile than do autos or railroads, and five times as much as buses. Surprisingly, there is not as much difference between railroad and auto travel for intercity trips as might be supposed. In urban travel, autos are big energy users; shifting commuters from autos to mass transit systems could save much energy and also reduce pollution. An even bigger savings would be accomplished if a wholesale change in automobile size could be effected in the United States. If autos in this country were the same size as the typical European auto, only half the energy used at present would be required.[55]

Space heating is another large energy user. The National Bureau of Standards has estimated that more efficient insulation could reduce the energy requirement for heating by 40 or 50 percent.[56] However, actual economies from this direction will come only slowly, as old buildings are replaced by new ones.

---

[55] *Energy and the Future, op. cit.*, p. 137.

[56] G. A. Lincoln, "Energy Conservation," *Science*, April 13, 1973, pp. 155–161.

A recent trend toward the use of electricity for heating purposes has had a negative impact on overall energy efficiency. Earl Cook has worked out an example of the energy needed to heat 50 gallons of water in two comparison cases. In the first case, natural gas is burned in a power plant to make electricity (efficiency: 32 percent, including transmission losses) and then the electricity is used in a domestic hot water heater (efficiency: 100 percent). In the second case, the natural gas is burned to heat the water directly (efficiency: 62 percent). In the first case, 234,000 Btu are consumed; in the second, only 121,000 Btu are needed.[57]

Much of the recent increase in energy consumption has been caused by the rapid spread of air conditioning. Room air conditioners have been found to vary widely in effectiveness, from 4.7 to 12.2 Btu per watt-hour. Hirst and Moyers have estimated that if the effectiveness of air conditioners were increased from the current average of 6 Btu per watt-hour to an average of 10 Btu per watt-hour, the saving in energy in the United States would be equivalent to more than 7 million tons of coal per year. To put it another way, this would be a reduction of 1500 acres which would not have to be strip mined for coal each year.[58]

If the United States were to embark upon a serious, determined effort to reduce energy consumption, the results could be dramatic. Most energy demand projections are extrapolations of the past. In California, the Rand Corporation was asked by the state legislature to conduct a study to see what the impact upon projected electrical energy use would be if energy conservation measures were undertaken on a broad front. The Rand conclusions were that electrical energy demand under these conditions might be 50 percent less in 2000 than that in current predictions, requiring only 45 new plants in California instead of 127.[59] Although the Rand report has not been accepted by the energy industry as valid, it underscores the importance that energy conservation can have for us in confronting our energy problems.

---

[57] E. Cook, "The Flow of Energy in an Industrial Society," *Scientific American,* September 1971, p. 135ff.

[58] Hirst and Moyers, *Science, op. cit.*, p, 1302.

[59] *Energy and the Future, op. cit.*, p. 144.

FOUR

Is engineering really a profession?

The self-consciousness conveyed by the above heading is a quality that is typical of engineering: the profession is exceedingly introspective. It is continually asking itself what it is, where it is going, and why it is going there. During the twentieth century, there have been no less than eight major studies in the United States concerning the education of engineers.

Science and engineering

The following saying received so much circulation during the late 1950s and early 1960s that it eventually almost assumed the status of a cliché:

> Every rocket-firing that is successful is hailed as a *scientific achievement;* every one that isn't is regarded as an *engineering failure.*

Some engineers became so sensitized to the foregoing statement that they winced every time it was made, because of the image it evokes of an engineering profession standing petulantly aside while the scientists accept all the glory.

If this were the statement's only significance, it could have been placed in the "file-it-and-forget-it" department, while we pass on to more important things; but, basically, the little epigram cited here dramatizes the confusion in which we have been foundering, concerning what is "science" and what is "engineering." Ignoring the issue of who gets the credit for rocket firings, it is important that we clarify our thoughts about these two words because the distinction is central to a man's self-image of what he is doing, why he is doing it, and for whom he is doing it.[1] If a man is working in an environment that demands an

---

[1] J. A. Stratton, president of the Massachusetts Institute of Technology, commented thus: "It matters not a whit whether a man doing a particular piece of work calls himself a physicist or an electrical engineer. But compre-

engineering result and if he instead produces a scientific one, it is probable that disturbances will occur. On a national scale, the effects could be important.

Most of the current definitions of these two words, "science" and "engineering," shed little light on the matter. In attempting to invent definitions that are acceptable to everyone, one is frequently driven toward the use of more and more general language, with the result that the definitions become increasingly vague. Soon everybody is satisfied, because anyone can read into the definition whatever he wants to. Thus, the definition is useless. Let us, therefore, use simple definitions and focus our attention upon the concepts behind them. For instance:

Scientists primarily produce *knowledge*.
Engineers primarily produce *things*.[2]

Note, however, that a given individual may at one time be producing knowledge, and at another, be producing things. This is because engineering and science are so intimately related. In fact, as our society grows technologically more sophisticated, engineering projects tend to become a mixture of these two: in order to achieve the final engineering objective of a project (the creation of a working device, system, or structure), it may be necessary to uncover much scientific knowledge which was previously unknown.

The basic task of the scientist is to perform *research* (create new knowledge), while the basic task of the engineer is to perform *design and development* (create new things). A number of specialized words have been used in this context: for example, it is said that scientists characteristically employ induction, while engineers employ deduction. (Induction is the process by which generalities are drawn from specifics, and deduction is the reverse.) Although there is nothing wrong with these terms, they have little concrete meaning for most people. Consequently, they are not so starkly graphic or useful as the simple terms with which we started. M. P. O'Brien has suggested that much of our confusion about engineering has resulted from the tendency toward "inflation" of terms; he writes: ". . . drafting is called design; design is

---

hension of the fact that physics and engineering are indeed different, with different professional missions, is essential,"—from "Physics and Engineering in a Free Society," *Physics Today*, March 1961, p. 23.

[2] Eric Walker, president of The Pennsylvania State University, has essentially stated the definitions set forth here, in his article "Engineers and/or Scientists"; he says: "Science aims at the discovery, verification, and organization of fact and information . . . engineering is fundamentally committed to the translation of scientific facts and information to create machines, structures, materials, processes, and the like that can be used by men."—from *Journal of Engineering Education*, February 1961, pp. 419–421. (For some dictionary definitions, see Appendix.)

called development; development is called research; and true research must be modified for clarity and called 'basic research.' "[3] (Some precise definitions of these terms have been developed by the National Science Foundation and are included in Chapter 12, on design and development.)

A scientist may be primarily a theoretician (that is, he may specialize in creating new theoretical explanations for previously unexplained phenomena), or he may be primarily an experimentalist. In either event, his final output is a report or paper published in a scholarly journal. As soon as a given bit of work is published, the scientist's job is complete with respect to that particular piece of information. In the ensuing discussion, the words "science" and "research" will be treated essentially as synonyms.

Engineers, on the other hand, are concerned with the creation of devices, systems, and structures for human use. However, it should be noted that the output of an engineer may not always be tangible. For example, many engineers are engaged in the design of intangibles, such as processes and systems, but in this discussion, such intangibles will be considered "things." Engineers also are frequently employed in liaison or consulting capacities in construction, testing, manufacturing, and as agents for governmental bodies. Such men are not directly engaged in design, but design is still at the core of their activities. Their role is to interpret the design and to see that it is carried out correctly.

It is true that, in many instances, persons originally trained as physicists or chemists have had to step in and handle important engineering tasks, probably because engineering education in the past generally did not carry the student deeply enough into science and mathematics.[4] Still, a person whose entire educational experience has been structured around a core consisting of research is likely to consider it a misdirection of his interests if he is expected to produce working devices or physical structures—hardware, as it is called. He is likely to believe that his proper mark of achievement is publication in a scholarly journal where his research contribution can be judged by acknowledged authorities in the field.

This discussion now leads to a consideration of the term "engineering scientist." Many persons in engineering believe that the term should not be used at all and that the words "engineering" and "science" are

---

[3] M. P. O'Brien, "Technological Planning and Misplanning," in *Technological Planning on the Corporate Level* (Cambridge, Mass.: Graduate School of Business Administration, Harvard University, 1962), p. 73.

[4] An excellent example is the development of radar at MIT during World War II. The participants readily admit that this was an engineering development task; yet over half of the 1000 top-level participants were basic scientists, mostly physicists. See "Longhairs and Short Waves." *Fortune*, November 1945, p. 163 ff.

mutually exclusive. This is one of the deeply controversial subjects in engineering. The following fact must be borne in mind, however: many science fields that were once considered the property of physicists or chemists have, in the last 30 years, become primarily the domain of engineers. These fields, designated as the *engineering sciences* by a committee of the American Society for Engineering Education (ASEE), are 1) mechanics of solids, 2) merchants of fluids, 3) transfer and rate processes, 4) thermodynamics, 5) electrical sciences, and 6) nature and properties of materials.[5] The difference between engineering science and any other kind of science is that engineering science seeks new knowledge for the specific purpose of facilitating the design and development process, while *science* in general seeks knowledge without regard to its application. The latter is sometimes called "basic" science.

In 1962, the President's Science Advisory Committee published a report that has had a deep impact upon the engineering profession: the report called for a massive build-up in the number of graduate degrees in engineering, mathematics, and the physical sciences to meet the future needs of the nation. The Committee recommended that some of these men prepare themselves for research and others for creative design.[6]

Who is an engineer?

One of the difficulties in establishing who is, and who is not, an engineer is that there are no clear-cut guide lines. For example, it has been estimated that 38 percent of the 1.2 million engineers in the United States in 1972 did not have college degrees, that only about 25 percent were legally registered, and that only 29 percent were members of one or more of the five major societies known as the Founder Societies.[7] With such a heterogeneous engineering population, it is easy to see why many people honestly ask the question, "Is engineering really a profession?"

There is a great attractiveness to the title of "engineer." Many groups (and individuals) have sought the right to use the title, and those who have historically possessed this right have strived hard to

---

[5] *Report on the Engineering Sciences* (Washington, D.C.: ASEE, 1956–1958).

[6] *Meeting Manpower Needs in Science and Technology, Report No. 1: Graduate Training in Engineering, Mathematics, and Physical Sciences* (Washington, D.C.: The White House, December 12, 1962).

[7] *Selected Characteristics of Five Engineering and Scientific Occupational Groups, 1972* (Washington, D.C.: National Science Foundation, July 20, 1973). *40th Annual Report, 1971–72* (New York: Engineers Council for Professional Development, September 30, 1972). (For description of Founder Societies, see Chapter 14 of this text.)

keep it. For example, locomotive engineers and stationary engineers (power plant operators) wish to preserve their traditional names, and it must be granted that they probably have as much historically established right as anyone to use these titles.

Many "subprofessional" and semiprofessional people have sought the title of engineer and generally mark it a red-letter day in their lives when (and if) that goal is achieved. As an interesting sidelight on human behavior, it can be mentioned that previously sport-shirted technicians have been observed to pop up in white shirt, tie, and dark suit the day following a promotion to *engineer*.

Management men who have been basically educated as engineers frequently continue to think of themselves as the latter. Engineering societies are heavily populated by managers who are not currently functioning as engineers, but who continue to be active in engineering affairs. Charles Wilson, Secretary of Defense under President Eisenhower, and prior to that, chairman of the board of General Motors, was regarded by much of the public as an engineer, although he worked as an engineer for a relatively short time before starting the climb up the ladder of general management. Herbert Hoover, known throughout his long life as "The Great Engineer," actually was graduated from college as a geologist, having switched to geology from mechanical engineering in his sophomore year at Stanford University. He worked for a time as a field geologist but was quickly swept into world-wide mining activities, first as a mine manager, and then as an international businessman and financier with his operations based in London.[8] Even though only a small part of Mr. Hoover's career included direct engineering activities, apparently he was very willing to be known to the public as an engineer.

Public image

There are at least four classic sources of confusion concerning the question, "Who is an engineer?": 1) the matter of science versus engineering; 2) men who perform as engineers, but possess none of the usual hallmarks of the engineer such as college degree, registration, or professional society membership; 3) men who were originally educated as engineers, but who have actually done engineering for only a short time (or not at all) before making their careers in some other field such as sales, manufacturing, or management; and 4) the situation illustrated by the child who thinks an engineer is a man who drives a train. Even after these classic sources are removed from the picture, however, the public remains in a state of considerable confusion regard-

---

[8] H. G. Warren, *Herbert Hoover and the Great Depression* (New York: Oxford University Press, 1959), pp. 20–22.

ing what an engineer is and does. The technician was in no doubt: the engineer was a clean-cut fellow in a white shirt and a dark suit. But to some of the public, the engineer is a man in a dirty shirt and a hard hat.

The mass entertainment media—television, for example—presents one picture of how engineers appear to the other fellow. On one musical variety show, the script called for one of the group to exclaim, "Here come the engineers!" And onto the scene they came, complete with hard hats and dirty shirts. Apparently, this particular scriptwriter's image of an engineer called for him to have both feet firmly planted in the mud, peering down the Burma Road through his transit.

Yet, the ubiquitous hard hat is appropriate for only about 5 percent of all engineers (mostly those in construction or mining); of these, only a portion have a need for wearing them, and, then, only a fraction of the time. Hard hats are also worn by architects, lawyers, and members of boards of directors when visiting on the job—without becoming identified with those occupations. The hard hat image gives the impression that an engineer is a construction superintendent. Engineers frequently do become construction superintendents, but this kind of activity is not the engineer's *characteristic* one, which is the creation of new devices, systems, structures, or processes.

Today's typical engineer is a college graduate (and there is increasing likelihood that he has an advanced degree) and is at home with science and mathematics. He is a resident of Suburbia and is a commuter. He is absent from home on business trips more often than one might expect, is alert to his environment, and is often more competent at self-expression than he is usually given credit for being. He is ambitious (though not feverishly so), generally believes he can advance only by "going into management," and may vaguely resent this alleged "fact."

His salary and standard of living frequently make him the envy of other white-collar workers.[9] His orientation, which is usually toward technical achievement as a criterion of excellence and often studiously sway from the usual status and success symbols, may mark him as a "queer duck" in the eyes of these same white-collar workers.

In a detailed analysis of a group of 100 mechanical engineers, a team of psychologists found the following traits to be those most frequently mentioned: easy-going, little friction in personal relations; emotionally stable; active, energetic; direct, straightforward; conscientious. At the bottom of the list, or missing altogether, were traits such as snobbishness, conceit, and arrogance. Although the psychological team deplored

---

[9] The average salary for engineers in 1971, including those without degrees, was $14,029. For "other salaried workers" it was $10,405. These data are from *Money Income in 1971 of Families and Persons in the United States* (Washington, D.C.: U.S. Bureau of the Census, Series P-60, No. 85, December 1972).

that the scope of engineers' cultural interests did not measure up to their intellectual potential, they stated, "While they are not smooth, they nevertheless usually make a favorable impression because of their transparent integrity and sincerity."[10]

The engineer's customary habitat is an office (perhaps shared with one or two colleagues) or the conference room. He may occasionally be found using the tools that are characteristic of the subprofessionals he supervises, and this partially accounts for some of the erroneous notions the public has about him. For example, if he is an electronic engineer, he may occasionally be found in the lab, twisting dials on instruments or making connections with a soldering iron as he debugs a circuit; if his sphere is mechanical, he may occasionally be found at a drafting board, performing a particularly difficult bit of layout work; and, yes, if he is a civil engineer, he may occasionally be found peering through a transit, with his feet in the mud. None of these situations is really characteristic of the engineer in his professional capacity. Being an eminently practical person, he employs these tools upon occasion when he believes he will get the job done faster or better that way. The *characteristic* activity of the engineer is one of intellectual effort, basically directed toward creative design.

Self-image

One of the most regrettable things the author ever heard was a remark made by an engineer that he had described himself to his son as a sort of white-collar plumber. Such remarks have a self-fulfilling quality: if a man believes himself to be a white-collar plumber, then that is probably what he will become.

Similarly, in surveys of engineering professionalism in industry, the interrogators will occasionally encounter an engineer who refers to himself as "just a peon" or as a "factory hand."[11] Unquestionably, some companies do tend to treat their engineers as a commodity, a practice that encourages the "peon" frame of mind. However, it is also certain that the man who thinks like a peon will be one.

A related finding in the survey just cited is of great interest: the interviewees were shown cards containing lists of various things

---

[10] R. Harrison D. Tomblen, and T. Jackson, "Profile of the Mechanical Engineer III: Personality," *Personnel Psychology*, vol. 8 (1955), (Cleveland: Personnel Research Inst., Western Reserve University), pp. 480–481.

[11] *Engineering Professionalism in Industry* (Washington, D.C.: Professional Engineers Conference Board for Industry, 1960), p. 23. In cooperation with the National Society of Professional Engineers. (The survey findings were based on personal interviews with 350 engineers who worked for well-known large companies.)

engineers can do to build professionalism. The engineers chose the following as the most important:

Becoming better communicators
Advancing in technical competence
Becoming better aware of how engineers fit in with the company as a whole

When asked what engineers actually do to build professionalism, between 60 and 80 percent of the engineers said that they do the three things listed. To answer to the same question, only 20 to 40 percent of the *managers* agreed that engineers do these things.[12]

It is often said that engineers do a poor job of expressing themselves by either the written or the spoken word, and engineers humbly accept this judgment. It is indisputable that they frequently cannot spell (in one engineering survey, the statement was made that ". . . for certain individuals the spelling was suggestive of the orthography of the fifth grade").[13] In the case of the spoken word, however, they may do a better job than is generally attributed to them.

Engineers tend to have a poor opinion of their status in the eyes of society. In one survey, two thirds of the engineers interviewed said they believed that engineering is not yet fully recognized by the public as a profession.[14] Yet, the results of a survey of 3880 Chicago households taken by the *Chicago Tribune* in 1956 (and subsequently published in Vance Packard's *The Status Seekers*) seem to indicate that engineers are far harder on themselves than is the general public. Among seven status groups, engineers wound up in Group 2, along with top executives of local firms, newspaper editors, doctors, lawyers, local judges, and college professors at prestige schools. In Group 1 were top executives of national firms, architects, medical specialists, federal judges, and stock brokers. Group 3 contained bank cashiers, junior executives, high-school teachers, and office supervisors, among others.[15]

In a 1972 survey, conducted by the Opinion Research Corporation, engineers were ranked third in prestige (tying with ministers) behind physicians and scientists. The occupations rating lower in prestige than engineers were, in order, lawyers, architects, bankers, accountants, and

---

[12] *Engineering Professionalism in Industry, op. cit.*

[13] R. Harrison, W. Hunt, and T. Jackson, "Profile of the Mechanical Engineer, II: Interests," *Personnel Psychology*, vol. 8 (1955), (Cleveland: Personnel Research Inst., Western Reserve University), p. 315 ff.

[14] *Career Satisfactions of Professional Engineers in Industry* (Washington, D.C.: Professional Engineers Conference Board for Industry), p. 33. In cooperation with the National Society of Professional Engineers.

[15] Vance Packard, *The Status Seekers* (New York: Pocketbooks, 1961), pp. 98–99.

businessmen. The 1972 survey showed a marked change from a survey in 1963, which ranked engineers sixth, behind physicians, scientists, ministers, lawyers, and architects.[16]

What is a profession?

In cases where a definition of the word "profession" has been prepared by the same body that is seeking professional recognition, the process may be considered somewhat suspect; as it has been said, "To choose a definition is to plead a cause. . . ."[17] It would therefore be best to turn to Webster's Unabridged for a reasonably general meaning of the word as it is interpreted by nonengineers.

> **profession:** A calling requiring *specialized knowledge* and often long and *intensive preparation* including instruction in skills and methods as well as in the scientific, historical, or scholarly principles underlying such skills and methods, maintaining by force of *organization* or concerted opinion *high standards of achievement and conduct*, and committing its members to *continued study* and to a kind of work which has for its prime purpose the rendering of a *public service*.[18]

The italics have been supplied by the author in order to make those items stand out that constitute the essence of the definition. The italicized items might be examined, one by one, to see how well engineering fits the definition of a profession. Engineering certainly requires *specialized knowledge* and *intensive preparation*, but the degree of preparation is not so great as that in medicine or law, two favorite professions with which engineers frequently compare their own. For many decades, engineering has required four years of preparation beyond high school for entrance into the profession, while medicine and law have required from seven to nine years.

There is no doubt that the engineering profession has a very strong *organizational structure*. (In this book, an entire chapter, entitled "Engineering Societies," has been given over to the subject.)

There may be some question whether engineering meets the tests of *high standards, continued study*, and *public service*. However, it should be noted that these three points relate more to the behavior of the indi-

---

[16] *Science Indicators, 1972* (Washington, D.C.: National Science Foundation, January 31, 1973), p. 97.

[17] *The Annals of the American Academy of Political Science*, January 1955. Attributed to C. L. Stevenson by M. L. Cogan, in "The Problem of Defining a Profession," p. 105.

[18] By permission from *Webster's Third New International Dictionary*, copyright 1966 by G. & C. Merriam Co., Publishers of the Merriam-Webster Dictionaries. Emphasis added.

vidual professional than to his group. The following paragraphs will examine these issues in detail.

HIGH STANDARDS OF ACHIEVEMENT AND CONDUCT. To the author's knowledge, the most earnest effort conducted to date to learn something about the actual degree of professionalism in engineering is a survey made by The Professional Engineers Conference Board of Industry. When engineers were asked what engineering professionalism meant to them, the three most frequent responses were:

1. Technical competence and skill: high standards of learning and ability
2. Prestige for the profession: stature; dignity; respect
3. Become more like lawyers and doctors: raise standards as in medicine and law; have something like the AMA or Bar Association[19]

It should be gratifying that technical competence is number one, but on the other hand, it ought to be embarrassing to engineers that the second and third responses relate to the *fruits* of professionalism and not at all to what we do to merit the status. In this survey, personal qualities such as creativity, responsibility, and ethical standards were far down on the list.

In the same survey, the engineers' *managers* placed the matters of competence, responsibility, and ethics at the top of their lists, while the aspects of prestige were well toward the bottom. The discrepancy in these perceptions is unfortunate, especially so in light of the fact that the major professional engineering societies also emphasize competence, responsibility, and ethics. But only 29 percent of the engineers in the United States are members of these societies. While the engineering *profession* may satisfy the criterion of "high standards," a legitimate question exists whether all engineers as individuals also satisfy the criterion.

CONTINUED STUDY. This is a difficult matter, for just exactly what does "continued study" mean? The man who scans the articles in the magazines that cross his desk will probably state that he engages in continued study and so will the man who is pursuing a crushing load of evening studies at a university. As a practical matter, it probably does not mean the latter. This may be acceptable for a man who is trying to upgrade himself (by getting a master's degree, for example), but it simply is not reasonable to expect a man to be away from home three or four evenings a week (plus whatever evenings his job may require) on a permanent basis, during a period when he presumably

[19] *Engineering Professionalism in Industry, op. cit.*

is also trying to raise a family. Some companies have adopted "released time" policies to make it possible for qualified employees to attend day-time university classes (early morning hours, for example) partly at company expense. This is a highly satisfactory plan, but it requires cooperation from the university in scheduling classes, as well as from industry.

For a minimum fulfillment of the criterion for continued study, a man must keep up on the current literature in his field. It is true that some of the literature in engineering is difficult to read. No doubt, many men excuse themselves from reading it on exactly these grounds, but, often, this does not prevent them from energetically asserting their claims to professionalism.

Some engineers have discovered that they can coast for a long time on what they learned in college and during the first few years of their professional life. But coasting, by definition, means going downhill. Many men remark upon their feelings of surprise when they pick up some of their own college work and examine it after a few years lapse of time. They cannot believe they really wrote it or that they once understood all that material. Through the years, the mind is relentlessly busy at forgetting things; meanwhile, knowledge increases and diversifies. Hence, continued study takes on a double importance: not only must a man pursue the expansion of knowledge, but he also must take action against the erosion of his own knowledge. It takes a considerable amount of running just to stay in the same place.

PUBLIC SERVICE. Some young engineers misunderstand what is meant by "public service" and assume it means they should offer their services to the public at no charge. However, this would be an unrealistically severe interpretation. The engineer's public service role is a real one, but is not so evident as a doctor's or lawyer's. Moreover, the fact that most engineers are employees of corporations renders their public service even less apparent. The effect of engineering upon the public's well-being is a vital one, nevertheless. We are more fully aware of this when we reflect that there is hardly anything we touch, use, or eat that has not been subjected to an engineer's influence at some stage in its development.

Admittedly, a corporation engineer might be hard pressed to identify a connection between his activities on any given day and a specific human benefit. To convince himself of his service role, however, he should imagine a situation where his activities, and those of all others in similar capacities throughout all society, suddenly vanish. It takes only a minimum of imagination to realize that such a disappearance would have a profound and deleterious effect upon the nation's productivity, together with all which that implies: fewer jobs, lowered national income, and general economic decline.

Of course, if all blue-collar (or white-collar) workers were suddenly wiped out, the economy would also suffer an abrupt decline. They render a public service, too; every productive worker does. The difference is that an engineer is engaged in the creation of things *that didn't exist before*. He should be conscious that his creative efforts—however small they may seem—have an immense influence upon the future, when added to those of all other engineers.

The preceding paragraphs should establish that engineering meets all the criteria set forth in Webster's definition of a profession. However, just because engineering is a profession, it does not necessarily follow that every engineer is automatically a professional. Every individual remains charged with the responsibility to meet the criteria on a personal basis. In the final summing up, the professional status of a group depends upon the individual behavior of its members.

Professionalism by legislation

It should be noted that the federal government, in the Taft-Hartley Act, defines "professionals" in a way that includes engineers. Furthermore, the title "Professional Engineer" is protected by law in all states but three. The definition of a professional employee given by the Taft-Hartley law is reproduced in this section. Engineers should read it carefully because, being the law, it governs the relationship of every engineer with his employer.

> The term "professional employee" means—(a) any employee engaged in work (i) predominantly intellectual and varied in character as opposed to routine mental, manual, mechanical, or physical work; (ii) involving the consistent exercise of discretion and judgment in its performance; (iii) of such a character that the output produced or the result accomplished cannot be standardized in relation to a given period of time; (iv) requiring a knowledge of an advanced type in a field of science or learning customarily acquired by a prolonged course of specialized intellectual instruction and study in an institution of higher learning or a hospital, as distinguished from a general academic education or from an apprenticeship or from training in the performance of routine mental, manual, or physical processes; or
>
> (b) any employee, who (i) has completed the course of specialized intellectual instruction and study described in clause (iv) of paragraph (a), and (ii) is performing related work under the supervision of a professional person to qualify himself to become a professional employee as defined in paragraph (a).[20]

---

[20] *Labor-Management Relations Act, 1947.* (Taft-Hartley Act), U.S. code, title 29, chap. 7, par. 152(12).

One result of this definition is that professional employees may not be included against their wishes in a union with nonprofessional employees. Another result is that professional employees are "exempt" from the provision of the law which makes it obligatory that employers pay time-and-a-half for overtime.[21] An "exempt" employee's work "cannot be standardized"; he is paid his salary, not by the hour, but for accomplishing a responsible task. In fact, he frequently receives no pay at all for overtime; he is expected to fulfill his responsibilities, however long this may take. The law also makes it possible for engineers to keep their own time records instead of being required to punch a time clock. (The law requires that accurate daily records of times in and times out be kept for nonexempt employees.)

In addition, the National Labor Relations Board has used certain general criteria as a guide in determining whether a particular group consists of professionals. If state registration is typically associated with the group's classification (as for engineers), this has been considered strong evidence that the work is of professional level. Also, a college degree has been increasingly emphasized as a criterion. The Board has refused to accord professional status to work being performed by some groups on the basis that most of the occupants of positions in the groups did not possess college degrees.[22]

Taking the opposite stand, articles appearing in some union periodicals have striven to convince engineers that they are *not* professionals. An officer of one of the AFL-CIO unions has given his opinion that "engineers are *workers*,"[23] and that engineering and production employees have common interests.[24] He issued a call to all engineers to join the American Federation of Technical Engineers (AFTE), under the protection of AFL-CIO. (A chapter dealing with engineers' unions is included later in this book.)

The problem of numbers

One of the fundamental facts of life for engineers is that they outnumber every other professional group except teachers. Table 4-1 presents statistics from the 1970 U.S. census.

---

[21] *Fair Labor Standards Act of 1938*, U.S. code, title 29, chap. 8. [See par. 207(a) for overtime clause and par. 213(a) for exemption of executive, administrative, and professional employees.]

[22] *The Engineer in Industry in the 1960's* (Washington, D.C.: National Society of Professional Engineers, 1961), pp. 55–58.

[23] R. M. Stephens, "Engineers are *Workers*," *AFL-CIO American Federationist*, November 1958.

[24] R. M. Stephens, "Helluva Engineer," *The I.U.D. Digest* (Industrial Union Dept., AFL-CIO), Winter 1957.

Table 4–1 Comparison of engineers with other professionals

Occupation	*Year*		*Increase, percent*
	1970	*1960*	
Teachers (elementary and secondary schools)	2,774,800	1,796,800	54
Engineers	1,230,500	871,400	41
Accountants	812,400	495,900	64
Teachers (college and university)	490,700	194,800	152
Physicians and surgeons	281,700	232,800	21
Lawyers and judges	273,500	218,300	25
Dentists	91,000	83,200	9
Architects	56,900	38,100	49

Source: *U. S. Census of the Population: 1970* (Washington, D.C.: vol. PC(1)-D1, 1973), pp. 1–718 to 1–719.

Moreover, the engineering profession is expected to require another 20 percent increase during the decade of the 1970s; it is estimated that, as a result, there will be a total of 1,500,000 engineers in this country by 1980.[25]

For a profession that is based upon individual creativity, such numbers have disturbing implications. Some people doubt that it will be possible for engineers to cultivate individuality or to maintain a sense of personal responsibility under these conditions. They call attention to the masses of technologists employed by some industries (such as aerospace) and despondently conclude that the game is already lost.

The role of the silver-lining seeker is not always an admired one, but it deserves an attempt. Examining the aerospace industry, for example, we ask ourselves, "Where do great achievements originate?" They come from somewhere, for satellites orbit the earth, and many men have successfully gone to the moon. Technical achievements do not merely spontaneously arise from the milling activity, as life was once thought to arise from mouldering humus. Unfortunately, we can rarely recognize technological progress, even when it is taking place under our nose. We perennially seem to expect that idea-generation must occur in huge quantum jumps, but always in some other place. The truth is that progress is born out of a ferment of frustration, setbacks, disappointments, and efforts; it is rarely recognizable when

[25] *The Demand for Engineers* (New York: Engineering Manpower Commission, August 1972).

viewed on the personal level. Added together, the infinitesimal improvements become a veritable avalanche. Remember that an idea cannot occur without an individual human mind for it to occur in.

Engineer shortage: fact or fancy?

Ever since World War II, the engineering profession has been plagued by alternating cries of surplus and shortage. The effects are visible in Figure 4-1. The U.S. Bureau of Labor Statistics (BLS) is presumed to have caused the first violent swing in the late 1940s by its warning of an impending surplus of engineers. In 1950, freshman enrollments fell to their postwar low, only to rise once more as the first "shortage" developed in response to the Korean conflict and the subsequent competition with Russia. Then, in 1957, following some defense industry cutbacks, various articles in the U.S. press announced that the "engineer shortage" was a thing of the past. *Fortune* gave prominent display to the situation in an article entitled "The Turn in the Engineer Market" (subtitled "The postwar 'shortage' of engineers has suddenly disappeared. The question is now whether it is gone for good"). However, in the body of the article, *Fortune* said that some company recruiters still were not able to find all the engineers they wanted and declared that the long-term demand for engineers would continue to be strong in order to keep productivity and living standards rising.[26]

The adverse publicity appearing in the press created another "shortage." Freshman engineering enrollments dropped from nearly 80,000 in 1957 to less than 65,000 in 1962. The only conclusion possible is that young men were seriously concerned about the stability of engineering employment.

In 1963, the Bureau of Labor Statistics, reinforced by the Engineering Manpower Commission (EMC), warned that further serious shortages lay ahead, if engineering enrollments did not increase. At about the same time, the requirements of the aerospace industry for engineering talent began to produce an overheated condition in the engineer/scientist job market. In 1966, the so-called *Engineer/Scientist Demand Index* hit a peak of 220 on a scale which defined the 1960 demand level as equal to 100. After hitting this all-time peak, the *Demand Index* started a five-year downward slide as the economy began to stagnate and widespread aerospace layoffs occurred. It finally hit its all-time low of about 35 in the winter of 1970–1971. By 1974, it had worked its way back up above the 150 level.[27]

---

[26] "The Turn in the Engineer Market," *Fortune*, December 1957, p. 150ff.

[27] *Engineer/Scientist Demand Index: Nine Year Report* (New York: Deutsch, Shea and Evans, Inc., March 1970). The index is based upon the volume of advertising for engineers and scientists in 23 major newspapers and 15 technical journals.

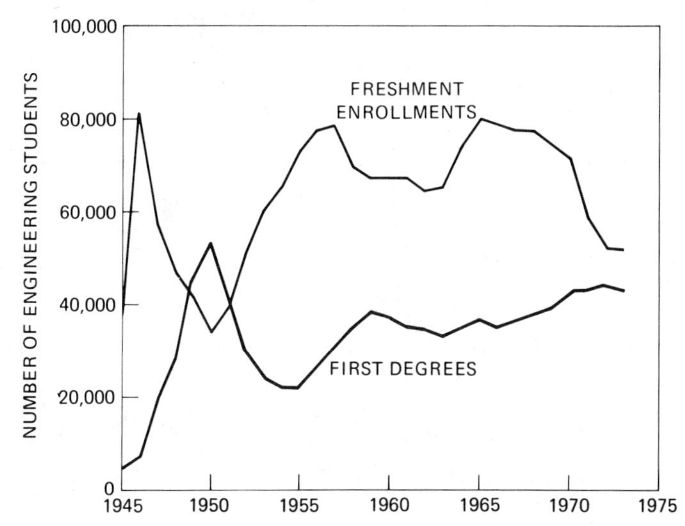

Fig. 4–1 Engineering freshman enrollments and first degrees since 1945. Source: H. Folk, *The Shortage of Scientists and Engineers.* Lexington, Mass.: D. C. Heath, 1970. Data after 1966 from Engineering Manpower Commission, New York, N.Y.)

One can argue about the validity of such demand indexes, but one cannot deny that important segments of the engineering profession went through deep waters in 1969–1971. Large numbers of engineers and scientists were laid off in aerospace and defense, and many had difficulty becoming employed again. The adverse publicity concerning such dislocations had a dramatic effect upon engineering freshman enrollments, which in 1972 dropped to their lowest level in 20 years; see Figure 4-1. Obviously, after a delayed reaction of four years or so, the supply of newly graduated engineers could also be expected to drop. This expectation led the Engineering Manpower Commission to predict that future supply would fail to meet the demand and create a "seller's market" for engineers. The EMC also pointed out that, even in the difficult years of 1971 and 1972, few engineering graduates were unable to find suitable jobs, implying that supply and demand for new graduates, even in those lean years, were about equal.[28]

These pronouncements of the EMC caused an immediate cry of anguish from some quarters. How could the EMC talk "shortage" when there were unemployed engineers around? Some professional engineers charged that the talk of shortage was only stimulated by engineering

[28] *The Demand for Engineers, op. cit.*, pp. 5–6.

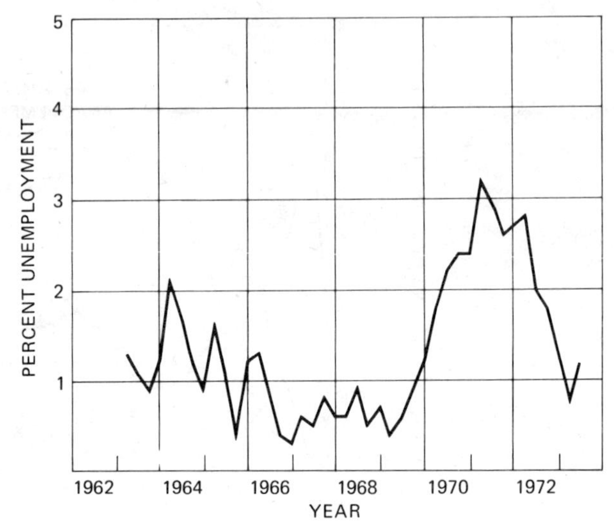

Fig. 4–2 Unemployment rates for engineers, 1963–1973. (Source: U.S. Bureau of Labor Statistics.)

deans anxious to fill their empty classrooms, and by company personnel men who would like to let expensive older men go and hire cheap new graduates. One critic declared that talk of shortages was totally self-serving and would cause a horrendous excess of engineering talent to be perpetuated.[29] However, at the very moment this particular critic was speaking of horrendous excesses of talent, the unemployment rate for engineers as reported by the U.S. Bureau of Labor Statistics, was standing at about 0.8 to 1.2 percent, only slightly higher than its level during the period of overheated demand in 1965–1968;[30] see Figure 4-2.

In 1971, a survey by the National Science Foundation (NSF) found that the peak of engineer unemployment showed a rate of 3.2 percent, about half the rate for all workers in the United States. However, if one were to add to the number of unemployed engineers those who were dissatisfied with their employment, either because they were in nonengineering work or were not working fulltime, the "employment problem" rate was 4.7 percent. The NSF survey produced some other

---

[29] *Wall Street Journal*, January 22, 1973. Letter to the Editor, from Chairman of Long Island Section, Institute of Electrical and Electronics Engineers.

[30] *Engineering Manpower Highlights*, Publication No. 225 (New York: Engineers Joint Council, November 1972). Also, *New Publications and Items of Manpower Interest* (New York: Engineers Joint Council, August 3, 1973).

interesting results. For example, it was learned that unemployment was related to educational background, as follows:

	Unemployment rate, 1971
Less than Bachelor's degree	4.3%
Graduate of nonengineering curriculum	3.7
Master's degree	3.2
Bachelor's degree	2.8
Doctor's degree	1.9

The survey also found that, contrary to popular impressions, the unemployment rate was not the highest among older engineers, but among those age 24 and under, where it stood at 5.6 percent. The rate for the 55 to 65 age group was about 4.2 percent, while for the group between age 25 and 54, the rate was 3.3 percent or less.[31]

The NSF survey of unemployment became the subject of some controversy. First of all, it covered only those engineers who were members of the 23 largest engineering societies, which constituted less than half of the nation's 1.2 million engineers. Second, the survey sampled one out of five persons in the selected group and had a response rate of 65 percent. Some critics said a large portion of those in the "no response" group were probably unemployed, and that it was like asking all those who were not present to please stand up. The National Science Foundation admitted that its figures could not be used on an absolute basis, but did have validity to show relationships with respect to age, educational background, and the like.

One of the more bizarre aspects of engineering unemployment was the publicity given to laid-off Ph.D.'s reduced to driving taxicabs or tending bar. No one knows just how many Ph.D.'s ever were in that condition, but it could not have been more than a handful. Nevertheless, the fact that these events had occurred at all was enough to make news. Simultaneously, there was national concern over the fact that new Ph.D.'s in some fields were not receiving employment offers. It soon became apparent that these reports mostly had reference to new Ph.D. holders in the sciences or humanities who had hoped to join prestigious universities. Since universities were no longer expanding, their aspirations generally could not be satisfied. Engineering Ph.D.'s were less subject to such conditions, because so many of them go into industry

[31] *Engineering Employment and Unemployment, 1971* (New York: Engineering Manpower Commission, October 1971).

instead of into teaching. A survey conducted by the National Research Council in 1971 showed that recent Ph.D.'s in engineering were having no special employment difficulties, with an unemployment rate of only 0.4 percent among 1969 graduates, and 1.5 percent among 1970 graduates.[32]

Many attempts have been made to explain why so many engineers remained unemployed as long as they did during the 1970–1973 period. The initial cause of unemployment was clear enough: it was the winding-down of the gigantic aerospace endeavor of the sixties after the moon landings had been accomplished. The number of scientists and engineers in aerospace declined from a peak of 235,000 in 1967 to 161,000 in 1973.[33] No doubt most displaced engineers found new employment readily, but many did not. The principal reason for their problem has generally been attributed to overspecialization. Aerospace pays high salaries and finds it necessary to have many of its engineers specialize deeply in narrow fields. If such a man is laid off, it is understandable that another company would be unwilling to pay a high salary for a speciality it does not need. Another problem was that many companies in the civilian sector were unwilling to take on engineers from aerospace or defense, for fear they were not sufficiently cost-oriented. Finally, many a laid-off engineer had a new job offered to him, but which required him to move his family and to accept a lesser salary than he had been making. It is not surprising that some in such straits preferred to remain unemployed in the hope something better would come along.

After the aerospace readjustments were substantially over and engineering seemed to have settled back to its accustomed state of virtually no unemployment, many asked what could be done to prevent a similar occurrence in the future. Obviously, the aerospace build-up had caused the overheated job market of 1965–1968, and the winding-down of that program in 1969–1971, coinciding as it did with a serious recession, had been the principal cause of unemployment. The fact that the wind-down obviously had to take place, and could have been foreseen, did not seem to satisfy anybody. But the lesson was clear. Any massive federal program of the future, undertaken in an atmosphere of crisis, could be depended upon to produce a repeat performance of the aerospace roller coaster ride: up—and then down. The federal government could hardly afford to deal this casually with such a valuable resource as its engineering and science manpower, and the consensus seemed to be that the government should assume the role of stabilizer and avoid massive crisislike programs if possible.

---

[32] *Employment of New Ph.D.'s and Postdoctorals in 1971* (Washington, D.C.: National Research Council, August 1971), p. 4.

[33] *Newsletter: Engineering for Industry* (National Society of Professional Engineers, August 1973).

The best summation of the technical manpower scene has been provided for us by Betty Vetter, Executive Director of the Scientific Manpower Commission. During the 1970s, she says, ". . . we will still be trying to erase urban blight, produce adequate clean energy, purge the environment, create effective transportation systems, and provide adequate health care, while maintaining our national defense and continuing some level of space exploration." Technical manpower will be vital to solving these problems, she reminds us, and consternation over displaced scientists and engineers should not be allowed to diminish our future ability to meet our needs.[34]

Professional ferment

As a consequence of the employment turmoil during 1969–1971, engineers began to demand that the various professional societies take a stronger interest in the individual engineer's personal welfare. Some of the societies established centers to aid in relocation and helped to set up retraining programs. A strong interest arose in sponsoring so-called "portable" pension plans in which an engineer could carry his pension benefits with him from job to job, the way many teachers do. Otherwise, if a man terminated a job, he lost all his pension benefits.

Some voices in the profession spoke strongly in favor of developing more protectionist societies and claimed that engineers should have an organization like the American Medical Association (AMA), which could control the supply of new graduates and presumably would have the ultimate effect of raising the salary for a senior engineer to the $35,000 level.[35] (Ironically, just at the time these voices were proposing that engineers organize like the AMA, the new president of the AMA was worrying that his organization was declining in public esteem. "Almost daily," he said in his inaugural speech, "there are attacks on our methods of practice, on our methods of payment, even on our motives and life styles. . . ."[36]) It has been pointed out by many persons that the proposal to control the supply of engineers would be sure to fail. In past periods of shortage, industry has met its needs for engineers by upgrading nongraduates from jobs as draftsmen or technicians, and by encouraging immigration of foreign-born engineers—7400 of the latter in 1972.[37]

---

[34] B. Vetter, "A Bubble in the Educational Pipeline," *Science*, April 7, 1972, p. 9.

[35] J. L. Markwalter, "A Doctrine for Engineers—or, Can the Engineering Profession Solve Its Own Problems?" *Professional Engineer*, July 1973, pp. 15–24.

[36] *San Francisco Chronicle*, June 22, 1972, p. 1.

[37] J. Alden, "Engineer Manpower: A Supply/Demand Issue," *Professional Engineer*, November 1973, pp. 35–36.

FIVE

Engineers in industry

In 1969, 72 percent of the nation's engineers were employed in private industry, according to a survey by the Engineers Joint Council. The survey also revealed that 14 percent were in government, 7 percent in education, and 4 percent were self-employed; see Table 5-1.

The engineer is almost always an employee of someone else. This fact in itself makes the engineering profession distinctly different from the three classical professions of law, theology, and medicine. Since he is a professional and yet an employee, the engineer represents a special and perplexing problem in employee relations.

Problems of professionals

Some managers believe engineers show more loyalty to their profession than they do to their companies. One top manager is reported to have said, "Engineers do not think of themselves as working for the company, but have the attitude that the company is a laboratory operated for their own special benefit." In a survey, the editors of *Machine Design* sought reactions to the statement just quoted, from 1000 engineers in industry, half of them in supervisory-management positions.[1] Most of the nonsupervisory respondents disagreed with the statement; some did so vehemently. The supervisory engineers were more cautious: only about a third said they believed the statement is false. A large segment of both groups thought the statement is true of some engineers, but very few said it is true of the majority.

At this point, a comment seems to be in order upon the basic assumption that a man must be loyal *either* to his profession *or* to his company and that he cannot simultaneously be loyal to both. This assump-

---

[1] "Management Charges . . . Engineers Deny." *Machine Design*, January 7, 1965, pp. 102–104. Survey performed in cooperation with Princeton Creative Research, Inc., Princeton, N.J.

Table 5–1 Distribution of engineers in 1969 by type of employer

Type of employer	Percent
Private industry and business	72
Federal government and military	10
Education	7
State and local government	4
Self-employed	4
Other	3
	100

Source: *A Profile of the Engineering Profession* (New York: Engineers Joint Council, March 1971).

tion could be a major cause of misunderstandings between engineering and management. It might help to clarify these issues if we distinguish between loyalty to professional goals and the blind pursuit of selfish objectives. It is the latter kind of behavior to which management generally objects and which gives rise to the kind of statements that stimulated *Machine Design*'s survey. True professionalism would require that the interests of the client (in this case, the employer) be primary, subject to the condition that the welfare of society is not damaged, of course.

The newly graduated engineer is generally unaware of the kinds of crosscurrents described in the preceding paragraph. He is likely to be much more concerned with his own personal problems and is eager to prove himself and find his place in the industrial world. Considering the intense anticipation that builds up in the mind of the student as the change to industrial life approaches, it is not too surprising if some subsequent disillusionment occurs.

One factor that may make the transition awkward is this: the new engineer starts his first job with a salary that often exceeds the amount paid to nonengineering employees in the organization who have had many years of experience. Some people might take the stand that the new engineer should immediately start making significant contributions in order to justify his high salary, but this point of view would be inconsistent with the frequently recorded opinions of engineering supervisors that it takes one or two years for a man to become a fully producing member of the team.

Some serious attempts have been made to identify and analyze the problems of young professionals. In his book *Characteristics of Engineers and Scientists*, Lee Danielson reports the results of interviews with 367 engineers and scientists, conducted by the Bureau of Industrial Relations at the University of Michigan. Danielson found the following to be the most frequently mentioned problem areas of young professionals, in order of their frequency of mention:

1. Adjusting to company practices
2. Advancement slow or uncertain
3. Accepting routine jobs
4. Learning what is expected
5. Finding one's own niche
6. Unrealistic ambitions
7. Lack of initiative
8. Gaining social acceptance
9. Lack of specialized courses
10. Lack of recognition[2]

Problems 4, 5, and 8 undoubtedly would diminish with the simple passage of time. Problem 1 can be helped by company training programs (about which more will be said later). The remaining items are probably the most important ones on the list and are, also, the most difficult to deal with constructively.

The essence of some of these points is that the young engineer is simultaneously being criticized for not being ambitious enough (Problem 7) and for being too ambitious (Problem 6). Great things are expected of him because of his education; at the same time, he is told that he does not know enough to handle anything important and he is given routine tasks. As an example, comments from two of the respondents in Danielson's survey are given below:

> I think most of the people think they are more important than they are. This is the college build-up, once again. You are important eventually, but not immediately.
>
> Their biggest difficulty is having to start on a low plane.

Remarks of the same sort turn up, again and again, in surveys and commentaries. They occur so frequently that a person wonders if these traits are often discovered in new graduates because they are so confidently expected to be there. On the other hand, there must be some substance to these criticisms, since they are reported so often. Perhaps the colleges *are* partially to blame, as suggested by the following quotation from a statement by an aerospace executive:

> We feel there is a strong need for the college to provide students with a more realistic image of the engineer on the job. Young engineers come to us with their heads in the clouds, looking only at the glamour of

---

[2] L. E. Danielson, *Characteristics of Engineers and Scientists* (Ann Arbor, Mich.: University of Michigan, 1960), pp. 55–78.

engineering. They should also appreciate the need for pick and shovel work, to get their hands dirty, and be practical.[3]

To achieve the desired amount of humility, and yet avoid the wallows of torpidity, poses a considerable problem for the young engineer. One writer warns against ". . . producing the apathetic, uncreative, passive kind of employee whom most organizations seem to welcome at the outset but regret being saddled with at a later time."[4]

Enthusiasm is one of the most valuable assets a young engineer possesses. Yet, it is this very enthusiasm that might sometimes be interpreted as overaggressiveness and unrealistic expectations. If consistently rebuffed, enthusiasm can easily degenerate to a permanently low level. This would be a tragedy because enthusiasm is too precious a national resource to be smothered—and eventually lost—in this fashion. Colleges and universities can perform the service Mr. Herman implies by teaching their students a simple formula: one of the best ways to move out of routine assignments is to handle each of them with as much skill, dispatch, and enthusiasm as one has at his command. There are few, if any, better ways in which to come to the favorable attention of management.

The problem of slow promotions is something else. Probably no one is ever promoted as fast as he would like to be. Moreover, most young men very quickly learn that a complacent attitude toward promotions may make them come slower than would otherwise be the case. Many have heard the adage that "squeaky wheels get the grease" and are not unwilling to test its truth in practice. But they may also learn that management is generally unimpressed by any agitation for promotion unless it is coupled with proved ability and achievement.

All these factors create problems, but they are not unique to engineering: they are human problems, and probably are permanently with us. The engineer can improve the situation through a desire to do the best job possible plus a a reasonable amount of patience. On the company's part, more specific actions are required, and these will be dealt with in the next section.

Constructive action by employers

Although not all the fault for the existence of problems lies with the companies, there are more avenues for improvement open to companies than there are to individuals. Once the individual engineer has diligently equipped himself with all the right attitudes—patience, enthusi-

---

[3] I. L. Herman, *A Frontier Industry, and Personnel Development Implications* (A Report to the California Commission on Manpower, Automation, and Technology, December 11, 1964).

[4] E. H. Schein, "How to Break in the College Graduate," *Harvard Business Review,* November–December 1964, pp. 68–76.

asm, diligence, alertness, thoroughness—the next moves are up to his corporation.

The proposals advanced by some writers will be very hard for many managers to accept. For example, Danielson's prescription is that management assume the role of "helper" in order to maximize the contributions that the professionals can make.[5] In effect, he is suggesting that management treat itself as a service to the engineering function. However, he does not mean engineers should become arrogant and irresponsible and demand that management serve their wants. What Danielson does suggest is that both sides adopt the following viewpoint: the engineer's responsibility is to do a good job of engineering; management's responsibility is to do everything possible to help the engineer do that job.

One constructive point upon which everyone appears to be agreed (at least in principle) is that good communication must be maintained between engineering and management. Principally, this means the engineers must be supplied with information concerning company objectives, particularly those which affect engineering projects. There is a natural limit to the degree that such a program can be pursued, however. Much information about company objectives may also be precisely that kind of information that will aid and comfort the competition, if it should come into their hands. Management can hardly be blamed if it tends to hold back on such sensitive information, since it knows full well that engineers do quit the organization from time to time, and sometimes join competitors. Unfortunately, this sensitive information is the very kind that is of greatest interest. Hence, the goal of complete communication is never achieved, although with honest effort it can be approached.

Another important problem area is that of salaries,[6] reviews, and promotion. It is generally agreed salaries should reflect the contributions made by the engineers. This is by no means an easy task, and many companies do a less than adequate job in this area. In one survey involving 350 engineers and engineering managers, 88 percent of the engineers thought it was imperative for their companies to establish salary progressions that faithfully reflect engineers' contributions; when asked if their companies realistically followed such a policy, only 23 percent answered in the affirmative.[7]

Good facilities are also emphasized in virtually every list of recom-

---

[5] Danielson, *op. cit.*, p. 76. [Others have made proposals that are essentially similar—notably, Peter F. Drucker in "Management and the Professional Employee," *Harvard Business Review*, May–June 1952, pp. 84–90; and William H. Whyte, Jr., in *The Organization Man* (New York: Simon and Schuster, 1956.)]

[6] The subject of salaries is so important that Chapter Ten has been devoted to it.

[7] *Engineering Professionalism in Industry* (Washington, D.C.: The Professional Engineers Conference Board for Industry, 1960), pp. 35–37.

mendations. Facilities include such things as secretarial and clerical support, provision of technicians and draftsmen, equipment, telephones, and reasonably private quarters. The day of the giant "bullpen," with engineers stacked at desks ranged row on row, is nearly a thing of the past. Most companies have gone to considerable trouble and expense to provide semiprivate quarters for engineers, with perhaps two or three persons per office.

It is often assumed that the purpose of such accommodations is to generate high morale. This is only part of it, however, for there is not necessarily a correlation between morale and productivity: high morale —if it is based too much on comfort and pleasurable surroundings— can be remarkably sterile. From the stockholders' point of view, the only possible economic justification for providing a good environment for engineers is that such conditions enable them to produce more and better work.

The Engineers Joint Council (EJC), which represents engineering societies having a combined membership of close to 500,000 professionals, has adopted the following recommendations for improving management–engineering relations.

The EJC recommends that:

1. Management utilize the services of engineers more effectively and thereby afford them opportunity for advancement and economic improvement.
2. Management recognize its responsibility to make engineers feel that they are a part of management.
3. Management survey areas of communication, recognition, and salaries, and where found wanting, correct to conform with standards of professional practice.
4. The engineer take inventory of his services and his actions to make sure that he has a professional attitude toward his work.
5. Engineering societies establish and employ appropriate means to maintain high standards of ethical conduct for professional achievement.
6. Engineering societies encourage the professional development of their members and promote proper recognition of the profession.
7. Engineering educators emphasize the characteristics of the profession.[8]

Training programs

Probably no company will readily admit it does not have a training program, although the term "training program" covers an immense

---

[8] *Raising Professional Standards and Improving Employment Conditions for Engineers* (New York: EJC, 1956), p. 14.

spectrum. At one end of the spectrum are formal programs that combine work with periods of full-time study leading to advanced degrees. At the other is "on-the-job training," wherein the new employee is put directly to work and is taught by his supervisor as the need arises. Just why the latter type of approach should merit the title "training program" is unclear, since this is the way new employees have been broken in since time began.

Contrary as it may seem to expectations, on-the-job training appears to be more popular with new graduates than do many types of formal programs. A survey of more than 1000 participants in 26 different training programs showed that on-the-job programs far surpassed the more formal types in the esteem of the participants.[9] This survey did not include any programs leading to advanced engineering degrees, however. Mostly, the intent of the "formal" programs was to prepare young men for the ultimate assumption of managerial duties.

According to the National Industrial Conference Board, the differences between formal and on-the-job programs are approximately as follows:

FORMAL TRAINING. May last one, two, or even three years. Time is spent in several different departments, mostly observing. Written reports and examinations are included. The emphasis usually is on preparation for management.

ON-THE-JOB TRAINING. Duration of program is generally less than that of formal programs. The training usually focuses more on a particular job, at the expense of a broader orientation, and involves actual work participation.

One explanation offered for the greater popularity of on-the-job training is that the new graduate, saturated with formal classroom education, is eager to get started in actual productive work. The strongest criticisms of some "formal" programs were that they were not challenging enough, moved too slowly, or did not contain enough actual "doing."

Edgar H. Schein, professor of Industrial Management at the Massachusetts Institute of Technology, listed the following kinds of "induction strategies" used by organizations:

1. *Sink or Swim.* The new graduate is simply given a project and is judged by the outcome. If he is given little information to guide him, then he is partially judged by how good a job he does in structuring

---

[9] S. Habbe, *College Graduates Assess Their Company Training, Personnel Policy Study No. 188* (New York: National Industrial Conference Board, 1963), pp. 45–46.

his own assignment. This requires the new employee to take vaguely stated objectives and translate them into specific tasks that can be dealt with one by one.

2. *The "Upending" Experience.* The intent of this type of strategy is to jar the new employee loose from the presumed "impracticalities" he acquired in college and to confront him with the "realities" of industrial life. In one approach of this nature reported by Schein, each new engineer is given a special electric circuit, which violates several theoretical assumptions, to analyze. When the new man reports that the circuit will not work, he is shown that it not only does work, but also has been in commercial use by the company for several years. Chastened, he is then asked to find out *why* it works. When he finds he cannot do this, he becomes thoroughly depressed and is now considered "ready" to tackle his first real assignment.

3. *Training While Working.* This is the typical on-the-job type of induction program. The new man is given an assignment, commensurate with his experience, and carries it out under the close guidance of his supervisor.

4. *Working While Training.* The new man is considered to belong to a formal training program, but is given small projects involving real work. He may be rotated through several different departments during the course of the program. It is sometimes difficult to decide whether programs of this type should be classified as "on-the-job" or "formal."

5. *Full-Time Training.* These programs clearly belong in the "formal" category. They usually involve class work and rotational assignments that call for the trainee to observe the work being done by others; direct participation is minimal. Schein observes that some trainees criticize such observational activities as mostly meaningless or "Mickey Mouse."

6. *Integrative Strategies.* In approaches of this type, an attempt is made to adjust to the different needs of different trainees. In one such program, the new employees are given regular job assignments for a year and then are sent to a summer-long full-time university training program. A key feature of the initial assignments is that the supervisors have been specially selected and trained to be sensitive to the new man's problems. Some of these programs lead to advanced degrees.[10]

Of Schein's six "strategies," the first two ("sink or swim," and the "upending experience") would seem to show little appreciation for the proper objective of any inductive strategy, which should be to turn a new man into a productive employee as soon as possible. It should be mentioned, also, that Schein did not imply endorsement of these two strategies: he merely recorded them. Moreover, Schein argues that

---

[10] See E. H. Schein, "How to Break in the College Graduate," *op. cit.*

initial assignments should maximize responsibility to the highest degree possible, for the sake of both the man and the organization. Admittedly, there is risk in such an approach because the new man could fail an important assignment. However, there is much to gain by using men at their highest potential, and much to lose by using them at their lowest.

The "upending" experience, especially, has hidden dangers. No doubt the purpose of such an approach is to demonstrate to the new man that all knowledge is not contained in textbooks and that nature still insists on behaving the way she wants to, without necessarily feeling constrained to conform to the behavior prescribed by men. While the purpose may be worthwhile, an "upending" experience is a poor way to achieve it. As a result, the new man often abandons "theoretical" approaches in favor of "practical" ones. A further-reaching consequence is that he has thus unknowingly limited himself in his professional development at a time in our history when an increasing number of important problems are proving to be solvable only by people having an excellent command of theory.

Professional employment guidelines

One of the major outcomes of the professional turmoil which followed the aerospace cutbacks of the late sixties was the development of *Guidelines to Professional Employment for Engineers and Scientists.* (The *Guidelines* are printed in their entirety in the Appendix.) By 1973, 15 engineering societies had endorsed the *Guidelines.* The emergence of such a code is a major event in the history of the engineering profession and recognizes these facts: 1) most engineers are employees of someone else; 2) the usual codes of ethics apply primarily to engineers in private practice, and are not very helpful to those engineers in industry, government, and education.

The *Guidelines* contain far more injunctions for employers than for employees. In brief, the employee is expected to be loyal to his employer's objectives; safeguard the public welfare; avoid conflicts of interest; and pursue professional development programs. The employer should keep his professional employees informed of the organization's objectives and policies; establish equitable compensation plans; minimize new hirings during layoffs; provide for early vesting of pension rights; assist in professional development programs; provide timely notice in the event of termination; and assist in relocation efforts following termination.

The *Guidelines* are not legally binding, but they are expected to have a constructive impact upon employee–employer relationships. The presence in them of certain items are clues to some of the problems which beset aerospace engineers who suddenly found themselves out of work. For example, the emphasis upon early vesting of pension rights reflects the fact that many persons found, after several years of service with employers, that they had acquired little or no right to the contributions made by the employers to the employees' future pension funds. Many—perhaps most—pension plans provide that the employees' rights to employers' contributions become "vested" only after many years of service. Thus, if an employee were to change jobs every few years during his career, he could easily wind up with no pension except what federal social security might provide.

A concern which is somewhat related to the one just described is reflected in the guideline to limit new hiring during periods of layoff. This guideline was stimulated by the realization on the part of some engineers that the companies which had just laid them off were continuing to hire brand new engineering graduates. The companies explained that they had new needs which could not be met by the persons let go, but the newly unemployed engineers were more likely to believe that their former employers were trying to save money by hiring less expensive manpower. Some companies replied by saying that the desire to save money could not be a correct explanation, because new graduates could hardly be considered cheap, were certainly expensive to recruit, and generally could not be very productive for a year or two anyway. Perhaps understandably, such explanations were received with skepticism by those who had been laid off.

The *Guidelines* are subject to periodic review and updating by the sponsoring societies as experience is gained through their use. Even though any such code can be expected to have imperfections, its existence is an attempt to fill a void of understanding which has long prevailed in professional relationships with employers.

Women in engineering

Upon occasion, the observation has been made that the most discriminated-against minority group in the United States is the female sex. In engineering, there doubtless are some men who believe women have no business being engineers. Fortunately, such discrimination is waning.

Whether discrimination is the cause or not, relatively few American women are engineers. It has been estimated that less than 1 percent of the engineers in this country are women. In 1971, only about 2.5 percent of the entering engineering freshman class in college were women; at the senior level, only about 1 percent were women. About 350 engineering bachelor's degrees were awarded to women in 1971 in the United States out of a total of more than 40,000 degrees.[11]

Women have shown they are capable of performing very well as

---

[11] *Women in Engineering* (New York: Engineering Manpower Commission, May 1972).

engineers. It is probably not aptitude so much as custom, that deters women from engineering. Engineering education is strongly oriented in an occupational direction, and Western society generally expects that men will be the breadwinners. However, as a result of the general trend toward more women in business and professional careers of every kind, it can be expected that more women will enter engineering.

Surveys show that two thirds of the women engineers in the United States are concentrated in research, design, and development. Electrical, chemical, and civil engineering—in that order—are the most popular choices of women engineering students. However, women have entered virtually all engineering fields. In fact, up until the 1960s, the most popular field for women engineers was industrial engineering.[12] One of our best-known women engineers, Dr. Lillian M. Gilbreth (of "Cheaper by the Dozen" fame) was an industrial engineer.

In Schenectady, New York, Nancy D. Fitzroy is an engineer specializing in heat transfer for General Electric. In 1963, Mrs. Fitzroy was elected chairman of the Hudson-Mohawk Section of the American Society of Mechanical Engineers, the first woman engineer to be so honored by ASME. Her comments concerning women in engineering are: "Working with all male engineers seems to pose no more problems than it did in school. Engineers in general are some of the most even-tempered, steadiest people whom I have met. They treat you according to the caliber of engineer you are. But may I add that for a girl it helps to be just a little bit better. . . ."[13]

Several engineering colleges have women engineering professors on their faculties. Among such women are Dr. Irene Peden at the University of Washington, Dr. Irmgaard Flügge-Lotz of Stanford University, and Dr. Madeline Goulard of Purdue University. Dr. Goulard offers this advice to young women thinking of careers in engineering: "If a woman is going to work in predominantly male surroundings, she must above all be herself, accept herself, and not try to reject or suppress her femininity."

It has been noted that only about 4 percent of women engineers are in management positions, whereas about 30 percent of men engineers are either in general management or are managers of major departments; see Table 5-2. A partial explanation for this disproportionate representation may be found in the concentration of women in research and development activities, which classically are not very likely routes to management. Also, women may not feel as free to make the frequent changes of residence which are usually considered necessary if a person aspires to management. No doubt, some discrimination against women

---

[12] I. C. Peden, "Women in Engineering Careers," in *The World of Engineering*, J. R. Whinnery, Ed. (New York: McGraw-Hill, 1965), p. 277.

[13] J. Kotel, "The Ms. Factor in ASME," *Mechanical Engineering*, July 1973, pp. 9–21.

Table 5–2 Supervisory responsibilities of engineers in 1969

Supervisory responsibility	*Percent*
General management of organization	10
Management of major department, division, or program	20
Supervision of project or section	22
Supervision of team or unit	12
Indirect or staff supervision	18
No regular supervision given	18
	100

Source: *A Profile of the Engineering Profession* (New York: Engineers Joint Council, March 1971).

has also been at work. Alice Stoll, who is a mechanical engineer and head of the Biophysics Laboratory, Crew Systems Department, Department of the Navy, says, "It would be foolish to pretend that advancement for a woman is not somewhat more difficult than it is for a man in comparable circumstances." However, she also feels that much of the pioneering work in altering these circumstances is already done, and that opposition is gradually crumbling.[14]

Marjorie Townsend, who is an electrical engineer and a project manager for the National Aeronautics and Space Administration (NASA), reports little difficulty in being accepted in a managerial role. When asked if her colleagues ever looked askance at her, her reply was, "not askance so much as amazed most of the time." She has worked full time as an engineer since 1951, first for the Naval Research Laboratory and then for NASA. During this period of employment, she also found time to bear four sons.[15]

A worry expressed by some employers is that women may not prove to be permanent employees. It is feared that most of them will marry, have babies, and retire to home life at just the time they should be entering the most productive phases of their careers. The risk is undeniable; however, the counterargument has been presented that large numbers of young male employees do not prove to be permanent, either, but move to other employers after a few years.

To help answer some of the questions concerning engineering careers

---

[14] J. Kotel, *op. cit.*, in *Mechanical Engineering*, p. 14.

[15] "NASA Satellite Project: The Boss Is a Woman," *Science*, January 5, 1973, pp. 48–94.

for women, the Society of Women Engineers (organized in 1952) published the results of a survey of 600 women engineering graduates. The survey revealed the following information:

1. Eighty percent were married, with an average of two children; over half of these were married to engineers or scientists.
2. Fifty-three percent were employed, of which 81 percent were employed full time; most of those who were not employed had small children.
3. Seventy percent of the women working thought they had been given every opportunity for professional advancement.[16]

One significant fact must be emphasized: scarcely any field open to women pays them better than engineering. In 1971, the average starting salary for B.S.-level women engineers was $885 per month, compared to men engineers at $877. In certain other fields in 1971, women were receiving starting salaries as follows: accounting—$812; computer programming—$746; general business—$618; community and service work—$573; secretarial work—$465. Thus, in the words of the Engineering Manpower Commission, women were ". . . virtually aristocrats of their graduating class."[17]

Racial minorities in engineering

With the exception of orientals, members of racial minorities are extremely scarce in engineering. For years, young men and women of oriental ancestry have chosen careers in science or engineering, but it is only recently that blacks, Mexican/Americans, or native Americans have been doing likewise. In 1972, only 2.2 percent of the full-time undergraduate engineering students in the United States were black, with the proportion for other racial groups unknown.[18]

One of the basic reasons why young black people—as well as young people of other ethnic backgrounds—have been skeptical about career prospects in engineering is the difficulty of pointing out successful instances of those with similar background who have gone before them. However, matters are slowly changing, especially as governmental programs are established which require equitable recruitment, hiring, and promotion policies.

It has been estimated that less than 3 percent of line managers in

---

[16] *Preliminary Report of the Survey of Women Engineering Graduates* (New York: Society of Women Engineers, June 1964).

[17] *Women in Engineering, op. cit.*

[18] *Engineering and Technology Enrollment*, Fall 1972 (New York: Engineering Manpower Commission, March 1973).

industry are black. Even so, some blacks have moved into top corporate spots as corporate vice-presidents of major companies. The executive vice-president for a division of Burlington Industries is a black, and so is the engineering vice-president of a Chicago manufacturing company. Nevertheless, many blacks who have "made it" recall that they had to overcome many unusual obstacles along the way, especially before matters began to improve in the 1960s. The complaint voiced by one black manager is that, ". . . you must prove yourself to be a superblack to be accepted as equal."[19]

Engineering-related activities

It was observed, earlier in this book, that the characteristic activity of the engineer is *design*. However, there are other kinds of technical activity that have traditionally been regarded as part of the engineering scope, even though they are not involved with design; see Tables 5-3 and 5-4.

CONSTRUCTION. Many engineers go directly into construction activities and operate more as managers than as engineers. Nevertheless,

Table 5–3 Employment function of engineers in 1969

Employment function	*Percent*
Planning, directing	20
Design	18
Advising, consultation	11
Development	9
Research	8
Production, operation, maintenance	7
Sales, technical services	6
Teaching, training	5
Liaison, estimating, budgeting, purchasing	5
Construction, installation	4
Testing, quality control	4
Other	3
	100

Source: *A Profile of the Engineering Profession* (New York: Engineers Joint Council, March 1971).

[19] E. Holsendolph, "Black Executives in a Nearly All-White World," *Fortune*, September 1972, p. 140ff.

Table 5–4 Distribution of engineers in 1969 by field of engineering

Field of engineering	*Percent*
Electrical, electronic, and communications	17
Construction and civil engineering	16
Aircraft and space	11
Machinery and mechanical equipment	10
Chemicals and allied products	7
Metals and fabricated products	6
Petroleum	5
Utilities	5
Educational and information services	5
Computers	4
Transportation	3
Mining	2
Ordinance	2
Other	7
	100

Source: *A Profile of the Engineering Profession* (New York: Engineers Joint Council, March 1971).

because of the highly technical content of their jobs, these men generally need engineering backgrounds, think of themselves as engineers, and belong to engineering societies.

OPERATIONS. Substantially the same remarks could be made about many engineers in manufacturing or processing companies who fill combined technical-managerial positions. In this case, they are usually industrial or mechanical engineers. Industrial engineers, for example, are often concerned with the organization of work activities, as in time-and-motion studies or work-simplification studies. They may also be involved with machine-evaluation studies aimed at obtaining maximum efficiency from a process. In some organizations, especially those concerned with processing, engineers may not only have technical responsibility for an operational unit, but may also have supervisory responsibility for the men who operate the unit. Many mining engineers, for example, fit into this category. Men performing such operational functions probably act more as managers than as engineers but are very likely to think of themselves as the latter.

SALES ENGINEERING. Many engineers move into activities that clearly consist of straight sales work and thus completely lose contact with engineering activities. However, this is not what is generally meant by "sales engineering," which is a province truly intermediate between sales and engineering. In fact, sales engineering occasionally involves engineering design. Such opportunities normally arise in enterprises that sell and produce custom-designed systems. In a typical case of this nature, a fully operating system, put together from off-the-shelf components, may be offered in a way that is unique to the customer. In some instances, it may be necessary to include a special component that has not yet been designed. The sales engineer works with the customer and essentially makes the sale, but he also designs the system to meet the customer's needs and when necessary, works with his home engineering office to develop hitherto nonexistent components. Not only do such people usually regard themselves as engineers, but indeed their work does have considerable engineering content.

Engineers in private practice and in government

Engineers in private practice

The number of engineers in private practice is small compared with the total number of engineers. For example, in 1970, the Consulting Engineers Council and the American Institute of Consulting Engineers had memberships of only 2300 and 420, respectively.[1] Obviously, not all consultants are necessarily members of these two organizations, but the figures are significant when compared with the membership totals of the Institute of Electrical and Electronics Engineers and the American Society of Mechanical Engineers for the same year—155,000 and 65,000, respectively.[2]

It is erroneous to think of a consulting engineer as an individual who typically offers his services to the public for a fee, like a doctor. There are some who function like this, of course, especially among those who are just starting out as consultants. But the "consulting engineer" usually is an organization that hires engineers, architects, accountants, draftsmen, clerks, and men of similar skills. Some consulting services are very large indeed and hire hundreds of engineers of all kinds: chemical, civil, electrical, mechanical, and nuclear, among others.

It is unlikely that engineers who are employees of such firms would see their positions as much different from those of engineers in industry. It is only the principals—the partners or officers of such companies—who have problems substantially different from other engineers. The bulk of their problems are typical of any businessman and involve finances, sales, legal responsibilities, and personnel. Yet, in reality,

---

[1] *Directory of Engineering Societies* (New York: Engineers Joint Council, 1970).

[2] *Ibid.*

their position is different from that of other businessmen, because their business responsibilities are coupled with *personal* responsibilities for the services they hold out to the public. This accounts for the frequent organization of large consulting services into partnerships or proprietorships: in a partnership or in a proprietorship (that is, a firm with only one owner), each principal is directly responsible for the firm's activities, whereas there are limitations to the liabilities officers of a corporation can incur. As a result, the corporate form of organization for engineering concerns has been the subject of controversy: opponents of such a form have argued that a corporation cannot be legally held responsible for services that are essentially personal in nature. Generally, for corporations engaged in consulting, it is required at least that the officer directly responsible for the engineering work be a registered engineer.

Becoming a consultant

A person beginning consulting work is much more likely to fail because of a lack of business ability than because of a lack of technical ability. Virtually all consultants warn the prospective newcomer about "that depressing first year." Some even declare that the lean period is apt to be three years instead of one.[3]

Some of the things the beginning consultant may neglect relate to such ordinary business matters as accounting, collections, overhead, taxes, and insurance. *Overhead*, for example, includes many items often overlooked. Vacations, sick leaves, insurance, and social security taxes may add from 10 to 15 percent of the direct costs. Rent, supplies, telephone service, and secretarial service may come to as much as 30 or 40 percent of the direct costs. If there are more than six or eight employees, additional supervision may be required; this may add another 15 percent. Finally, there is the often belatedly recognized factor of *nonproductive time*, which involves stand-by, fill-in, and other lost time, and may add another 10 percent. Thus, the direct engineering costs may have to be increased by as much as from 65 to 80 percent of the original estimate, with no allowance made, as yet, for profit.[4]

Probably, not even a determination to work extra hours would compensate for a lack of good business ability and adequate financial reserves, because extra hours and consulting work appear to go hand in hand as the normal situation. One consultant writes,

> If you object to working long hours and if you intend to dismiss all the business problems from your mind when you leave the office, don't

---

[3] J. S. Ward, "Starting Your Own Consulting Practice," *Civil Engineering*, January 1965, pp. 53–55.

[4] A. J. Ryan, "Operating Your Practice," *American Engineer*, November 1955.

try to be a consulting engineer, for the problems will be with you 24 hours a day.[5]

The same writer has offered the following list of *minimum* qualifications for the aspiring consultant:

1. Appropriate education, including humanities
2. Engineering registration
3. Confidence in his professional ability
4. Broad prior experience in responsible discharge of engineering work
5. Business acumen
6. Financial reserves to last at least six months (other writers say a *year* or longer)
7. Ability to get along with people, especially clients and employees

The basic problem of the new consultant is simple: to acquire his first project, he is expected to demonstrate his competence by pointing to projects he has completed in the past. Given such circumstances, getting the first job could be understandably difficult. Yet, every consultant in business today has had to get past this barrier.

Some consultants have attained their start by quitting their jobs and taking one or more of their former employers' clients with them. However, besides being obviously severely frowned upon, such a practice is unethical. Other aspirants have associated themselves with an established consultant, as a junior partner. This is certainly ethical, but offers a difficulty in that the senior man has to be convinced he has something to gain by taking on a new partner.

Compensation for consulting engineers

The size of the fee the consulting engineer receives will depend not only upon the size of the job, but also upon what kinds of services he agrees to perform. For engineering jobs involving constructed works, his typical services include the following:[6]

1. Advice on feasibility*
2. Preliminary studies and cost estimates*
3. Collection of basic data: surveys, borings, traffic census, etc.*
4. Preparation of plans and specifications

---

[5] J. B. McGaughy, "So You Want to Open a Consulting Office—By Way of Qualifications," *American Engineer*, October 1955.

[6] The items marked with an asterisk usually are contracted for separately from the main job.

5. Assistance and advice in awarding construction contract
6. Interpretation of plans during construction
7. Checking shop drawings
8. Approval for payments to contractor
9. Resident engineering service*
10. Inspection upon completion and supervision of tests
11. Preparation of final record drawings
12. Assistance during start-up
13. Consultation as needed[7]

In the past, the engineer's fee was often calculated as a percentage of the final construction cost, but there now appears to be a trend away from this. One of the arguments against this historical practice is that it penalizes the engineer for producing an economical design, when he should actually be rewarded.

Following are a few of the methods employed today for establishing the consultant's fee:

LUMP SUM. When the services to be performed are known with considerable precision, it may be possible to agree upon a fixed sum as the engineer's compensation. The obvious disadvantage is that the consultant may incur serious loss if the job has been underestimated.

PAYROLL COST TIMES A MULTIPLIER. Under this method, the client essentially pays the engineering costs as they occur, including a sufficient amount to cover overhead and profit.

PER DIEM. If the job is a short-term one, a fixed daily rate may be charged; this is known as a *per diem* arrangement. Direct out-of-pocket expenses, such as travel costs, are reimbursed in addition to the per diem payments.

PERCENTAGE OF CONSTRUCTION COST. Even though the "percentage of construction cost" method is not so widely used as it once was, such a calculation is often used to check the overall amount arrived at by means of one of the other methods. Charts in general use by some of the professional societies indicate that the engineering fee might be $80,000 to $100,000 for a million-dollar construction job and $500,000 to $600,000 for a 10-million-dollar job.

Many kinds of consulting engineering services do not involve construction. For example, some consulting concerns make a business of

---

[7] *Guide for Selecting, Retaining, and Compensating Professional Engineers in Private Practice* (Washington, D.C.: National Society of Professional Engineers, 1963).

performing product-development services for clients. Often, firms of this type, upon request, will also place their own personnel within a client's firm, to work side by side with the client's engineers. In this fashion, the client can absorb unexpected peak workloads without hiring and training people who may become surplus when the peak has passed. Consulting companies of the type just described usually charge a fixed rate per hour per person employed on the project, for as long as the client employs their services. The rates are set high enough to enable the consulting company to recover all costs, including overhead, and to provide it with a profit.

Ethical problems in consulting practice

A special problem that besets engineers in private practice has to do with competitive bidding. Bidding on the basis of price alone is held to be unethical. To persons who have been raised in an economic society founded upon competition, an ethical sanction against competitive bidding may seem eccentric. However, consulting engineers point out that people do not ordinarily select a doctor or a lawyer on the basis of price alone. They assert that the personal nature of the services offered by an engineer places him in a similar position.

The American Institute of Consulting Engineers' position on competitive bidding has been aptly phrased by its president, F. S. Friel:

> Competitive bidding for engineering services is not in the public interest since it may lead to the employment of the engineer least qualified for the particular work under consideration instead of the best qualified, which should be the objective.[8]

As an alternative to competitive bidding, the Coordinating Committee on Relations of Engineers in Private Practice with Government[9] recommended the following procedure:

1. Examine the qualifications of several potential engineering firms who may have the desired capabilities.
2. Select up to six firms with acceptable qualifications and investigate them carefully, including the use of personal interviews.
3. Rank the firms in the order of their desirability and commence negotiations with the top-ranked firm.

---

[8] F. S. Friel, "Ethical Problems," *American Engineer*, December 1955.

[9] The Coordinating Committee is composed of representatives from the American Institute of Consulting Engineers, American Road Builders Association, American Society of Civil Engineers, Consulting Engineers Council, National Society of Professional Engineers, Engineers Joint Council. (The EJC is an observer.)

4. If the negotiations are not successful, then open negotiations with the second-ranked firm, then the third, and so on, until negotiations meet with success.[10]

Much controversy over bidding on contracts by engineers exists within the profession. Some codes of ethics retain clauses condemning the practice, while other codes, such as the one published by the Engineers' Council for Professional Development (see Appendix), do not mention bidding.

Another unusual problem of consulting engineers involves their relationships with architects. Both groups are legally entitled to offer somewhat similar services; however, it is also recognized that each group possesses special competence in certain areas and that it is to their mutual advantage to work together. Furthermore, the public benefits by such cooperation.

Unfortunately, the relationships between these two professions have not always been smooth. As an example, a headline in *Engineering News-Record* for July 9, 1964, read, "NSPE Irked by AIA Ethical Standard."[11] The standard in question appeared to prohibit architects from working as employees of engineers. In replying to the article, the AIA claimed its standards had been misinterpreted.[12] The institute explained that its objection primarily focused on architectural-engineering concerns that employed registered architects only as minor employees. Architectural services should be performed within an established architectural division of the company, under the responsible direction and control of a registered architect, said the AIA.

Legal responsibilities

In the course of carrying out a contract, an engineer can sometimes acquire unexpected and unwanted legal obligations. "Supervision" and "inspection," two words frequently used in engineering contracts, frequently also cause trouble.

SUPERVISION. Many contracts in the past have used the phrase, "the Engineer (or Architect) shall have general supervision and direction of

---

[10] *A Guide for the Selection of Engineers in Private Practice* (Washington, D.C.: National Society of Engineers, September 28, 1961).

[11] NSPE stands for National Society of Professional Engineers; AIA stands for American Institute of Architects.

[12] "AIA Clarifies its Ethical Standards," *Engineering News-Record*, July 30, 1964, p. 12.

the work." In some recent court decisions, it has been held that the design professionals (a term embracing both engineers and architects) were responsible for defective construction techniques in cases where they undertook the responsibility of supervising.[13] In 1961, a joint professional committee of architects and engineers recommended that all reference to "supervision" be omitted from contract documents. The matter remains controversial, however, for many consulting engineers believe that supervision of the work is an engineering service to which the owner is traditionally entitled.

INSPECTION. The work "inspection" has caused trouble because it has sometimes been interpreted to mean exhaustive and continuous inspection of all details of the construction. Most often, this has not been the type of function the engineer had in mind when he agreed to "inspection of the work." More likely, he envisioned some kind of educated spot-checking; hence, the word "observation" has been proposed as a substitute that more accurately describes the service intended. If actual detailed inspection is desired, then it is recommended by professional groups that the contract provide for a full-time project representative whose task it is to perform detailed and continuous inspections.

Special legal hazards are involved in the use of new materials or equipment. Courts have generally held that the engineer or architect is obliged to conduct tests of the new material or to have reliable information concerning the results of tests conducted by others. Sole reliance upon manufacturers' sales literature and specifications has been held insufficient. The question of the obligations of the design professional, in the use of new materials, remains a tricky legal matter.

The foregoing paragraphs describe some of the legal hazards involved in offering consulting services. Any one of them could be financially catastrophic for the consultant. Because of this, professional groups recommend that architects and engineers maintain professional liability insurance, often known as "errors and omissions" insurance. It is further pointed out that the written language of the contract may be insufficient to protect the engineer, if he attempts to perform services in an area beyond the scope of the contract. J. R. Clark says, "Having once moved into that area he may be charged with the responsibility for all of the functions involved, such as failure to exercise reasonable care in per-

---

[13] J. R. Clark, *Concerning Some Legal Responsibilities in the Practice of Architecture and Engineering* (Washington, D.C.: AIA, 1961). This publication is the source for most of the information contained in this section. (Mr. Clark is a partner in the well-known legal firm of Barnes, Dechert, Price, Meyers & Rhoads, of Philadelphia, Pa.)

forming the services or failing to do what one experienced in the field would do in the exercise of reasonable care."[14]

Engineers in government

Many engineers in the federal government, and some in state governments, are engaged in research, design, and development. As a result, they probably would not perceive their situation as being much different from that of engineers in industry. Yet, there are areas of engineering activity in government that have little or no counterparts in private industry; these contribute to the formation of public policy and law enforcement, as in environmental regulation, operation of public utilities, and public transportation.[15] While jobs in such areas require engineering backgrounds, they frequently do not involve the basic engineering function of design. Much of the typical activity of government engineers has to do with the preparation of functional specifications for public works and with supervision of the resulting construction and operations. In the federal government, engineers have participated at high levels in policy matters, since the government has increasingly come to recognize that many decisions of national importance rest primarily upon technical considerations.

The direct involvement of government engineers in design activities has led to complaints by some consulting engineer groups against the use of government engineers in the design of public works. The consultants maintain they can offer greater economy in performing engineering services for the various governments than can the engineers who are *in* government. However, professionals on the opposite side disagree. This dispute—which actually is only one facet of a larger problem, namely, which services will be performed by the public sector of the American economy and which by the private sector—will probably continue to be troublesome in years to come.

The federal government has recognized that it has just as much need for high-quality engineering talent as does private industry. It has also shown concern that its image is not so favorable as could be desired. As a result, the Federal Council for Science and Technology in 1962 published a series of recommendations for the improvement of the technological environment in government. Among other things, the council recommended that the federal government:

1. Provide for greater participation by scientists and engineers in the making of decisions

---

[14] J. R. Clark, *op. cit.*, p. 23.

[15] A. C. Stern, "When Government Hires an Engineer," *Mechanical Engineering*, June 1964, p. 22.

2. Delegate more administrative authority to technical directors
3. Provide a clearer picture of the opportunities and challenges in a government career
4. Improve recruiting procedures
5. Provide better fringe benefits[16]

---

[16] *The Competition for Quality* (Washington, D.C.: Federal Council for Science and Technology, April 1962).

SEVEN

Management

It is interesting that many engineering students do not really want to be engineers: they want to be managers. However, they are generally quick to admit that they do not have a particularly clear picture of what being a manager would be like.

When asked why they want to be managers, engineering students may have some interesting answers: "Because I'm mercenary" says one; this is merely a way of stating that the prospect of large financial rewards is one of the traditional attractions in management. "Because managers run things," says another; this man's answer shows that he is getting very close to the core of the matter. An especially interesting reply is, "Because I don't believe I have what it takes to be a good engineer." Nevertheless, most students take the view that nearly everyone wants to grow in stature throughout his career, and management is the traditional route for such growth.

What do managers really do?

There is a simple answer to the question of what managers do. It can be answered that they manage money, materials, and men. However, this is little help to the potential management aspirant. He wants to know what it would *feel* like to be a manager. Would he like it? Would he be engaged in a daily rapid-fire round of decision-making à la Hollywood? Would he be constantly embroiled in dirty political in-fighting, dramatized in "Executive Suite" and "The Carpetbaggers"? Would he be required to ride the social merry-go-round regardless of his own wishes? Would he find it necessary to discipline his wife, at intervals, regarding proper corporation-wife behavior? The answer need not necessarily be "yes" to *any* of these questions. However, any given individual could make it "yes" to all of them, depending upon the level of the job and upon his own characteristics.

Up to a certain level, a man can belong to himself. That is, he can be fully loyal to his company, have high devotion to duty, be on the firing line to meet every crisis, and, yet, never seriously be placed in a position where he must choose where his greater loyalty lies—with his family or with his company. However, if he truly aspires to a high-level post he should not have any delusions: it will be necessary for him to marry the job.

The editors of *Fortune* estimate a 57- to 60-hour average work-week for executives, including the effect of take-home work and business entertaining. Furthermore, *Fortune* suggests that, while corporations may publicly deplore the pressure placed upon their executives, they privately do everything possible to increase it. One executive is quoted as saying, "What it boils down to is this: You promote the guy who takes his problems home with him."[1]

A large part of any executive's time is taken up by reading letters, memos, company reports, policy statements, proposals, specifications, laws, regulations, contracts, analyses, magazines, and newspapers. In addition, he must write (mostly memos), although he has a dictation machine and a secretary to ease this part of the chore for him. Much of this reading and writing is done at home, because the executive's day is almost monopolized by meetings.

During the office day, the executive is seldom alone. If he is not involved in a formal committee meeting of some sort, he is informally engaged with one or more associates, because the executive is primarily concerned with *people*. The single most characteristic thing about an executive's job is that he must somehow be able to get numbers of people moving in the same direction; hopefully, he gets them to move on a willing basis.

Once a given corporate objective has been formulated, a manager is faced with the task of getting others to attain it. He may feel himself more capable of carrying out any given task than the person to whom he assigns it, but being only one man, he cannot do everything himself. Therefore, he uses all those techniques that, collectively, are in the area of human relations. Some of these are good; some are bad. The executive suggests, recommends, persuades, urges, or directs (all good); or he may order, command, demand, fume, bully, or rage (listed in order of their decreasing desirability).

J. Irwin Miller, chairman of the board of Cummins Engine Company, has graphically described the immense multiple pressures that operate on an executive:

> To illustrate, let us suppose we can see inside the head of the president of a large manufacturing organization. His company employs 20,000

---

[1] *The Executive Life* (Garden City, N.Y.: Doubleday, 1956), p. 65. By the editors of *Fortune*. This book, plus Vance Packard's *The Pyramid Climbers* and William Whyte's *The Organization Man*, should be required reading for every management aspirant.

persons and operates half a dozen plants. It distributes its products in every state and in many foreign countries, and—most frightening of all—it has competitors.

Now let us suppose that these competitors are extremely vigorous, and that our president knows that to maintain his share of the market and to make earnings which will please his directors, he must accomplish the following very quickly: design and perfect a brand-new and more advanced line of products; tool up these products in such a way as to permit higher quality and lower costs than his competitors; purchase new machinery; arrange major additional long-term financing. At the same time his corporation's labor contract is up for negotiation, and this must be rewritten in such a way as to obtain good employee response and yet make no more concessions than do his competitors. Sales coverage of all customers has to be intensified, and sales costs reduced. Every one of these objectives must be accomplished simultaneously, and ahead of similar efforts on the part of his competitors—or the future of his company is in great danger. Every head of a corporation lives every day with the awareness that it is quite possible to go broke. At the same time he lives with the awareness that he cannot personally accomplish a single one of these vital objectives. The actual work will have to be accomplished by numerous individuals, some actually unknown to him, most of them many layers removed from his direct influence in the organization.[2]

The second most characteristic thing about a manager's job is decision-making. As we will show later, the engineer is a decision-maker, too, and by virtue of this function, is also a manager; however, the effects of decisions made by a person directly in management are more immediately apparent than those made by an engineer. The effects of an engineer's decisions may ultimately be far-reaching, but they are usually subtle. In fact, the engineer may never become fully aware of the ultimate effects of his own actions. This is not the case with the executive, who may be acutely aware that his decision, today, to cut back production means that hundreds of people will be out of work tomorrow.

Making decisions is a lonely privilege. Inevitably, there comes a time in the career of every person who follows the management route, when he realizes there is no one to whom he can turn for help in making his decisions. Before this point, there always was a boss to whom he could go for advice. Suddenly, he realizes that he himself is the boss to whom *others* are coming for advice. Worse: in considering the possibility of going to his own boss for assistance, he may realize, with a shock, that his boss cannot help him. There are many possible reasons why this might be so: the boss may be too far away; he may not have the

---

[2] J. Irwin Miller, "The Dilemma of the Corporation Man," *Fortune*, August 1959, p. 103.

special background necessary to understand the situation; he may not have the necessary time—or the manager may fear his boss will not *take* the time. Finally, it may simply be that the boss has delegated major responsibilities to the manager and *expects* him to make decisions. This is complicated by the fact that one of the hardest decisions to make, is the decision of what to take to the boss and what to determine for one's self.

In the effort to ease this loneliness, executives may turn to committees to help them make decisions. It may even be possible for executives to bury the responsibility for their decisions in the anonymity of committee action. But, under the methods that govern American business, this is feasible only to a limited degree. Even if a decision has been made by a committee, the ultimate responsibility for a bad decision will probably come to rest upon the individual executive who originally had the responsibility for acting. Knowing this, most executives will use a committee in two major capacities: 1) as a sounding board to enlarge their own perceptions of the problem, and 2) as a means to involve their associates and subordinates in the decision-making process and, thus, to approach unity of purpose as nearly as is possible. Following a committee meeting during which the problem has been thoroughly aired, the executive will then make the decision himself.

Levels of management

So far, it would seem to be the case that "management" looks and feels the same wherever it is found. It does not: it varies considerably, depending upon the level of management. A list of some management levels, together with typical titles of the men in each category, follows:

EXECUTIVE. Chairman of the board, president, vice-president (of manufacturing, of engineering, of marketing, of finance, and so forth), general manager, treasurer, controller.
MANAGER. Plant manager, chief engineer, director of engineering, general sales manager, personnel manager.
SUPERINTENDENT. Chief project engineer, chief industrial engineer, purchasing agent, group head, regional sales manager, assembly-line superintendent.
SUPERVISOR. Project engineer, foreman, office manager.

In *The Executive Life*, the editors of *Fortune* list the five characteristic functions of an *excutive* as 1) setting policy, 2) making major decisions, 3) coordinating, 4) organizing, and 5) delegating responsibility. The term "top management" is frequently used to identify this level.

According to *Fortune*, a man at the *manager* level does not set policy, but interprets and carries out policies formulated by others. He may have the authority to make decisions of considerable importance, for instance, approval of union contracts, but he makes these decisions within the limits set by top-executive policy.

A *superintendent* differs from a manager primarily in the magnitude of the decisions he may make and, from the category beneath him, by virtue of his function as a *supervisor of supervisors*. The term "middle management" generally encompasses the two categories of manager and superintendent.

A *supervisor* enforces rules; sees that quotas are met; administers personnel matters; and, in other ways, operates within fairly narrow and well-defined limits. Usually the term "first-line supervisor" is employed to describe his activities; this phrase signifies that the people under him are the productive workers themselves. (It should be noted that the "project engineer" is included in this lowest category of management. On the other hand, as is explained in this chapter, the project engineer actually is in a position of critical importance and may be considered a member of one of the most influential classes of people in industry. However, the purely supervisorial portion of his activities correctly belongs at this level.)

Generally, when a person says he wants to go into "management," he means he wants to be an "executive" or a "manager"; he is likely to regard the positions of supervisor or superintendent as way-stations on the road to ambition's fulfillment. Most of the discussion herein will be relevant to these two higher categories, since these are the levels where money, prestige, and influence (all highly desired) and stress, politics, and loss of personal freedoms (deterrents) are most in evidence.

Theories of management

So many different theories of management have been developed, that one set of authors has referred to the situation as "the management theory jungle." Nevertheless, the various theories do represent substantial differences in approaches to management. These theories or "schools" of management have been grouped by Koontz and O'Donnell briefly as follows:[3]

The operational approach, sometimes referred to as the "traditional" approach, looks at management by analyzing its various functions: planning, organizing, staffing, directing, and controlling. The adherents

---

[3] H. Koontz and C. O'Donnell, *Principles of Management: An Analysis of Managerial Functions*, 5th ed. (New York: McGraw-Hill, 1972), pp. 1, 34–42. Koontz and O'Donnell define "management" as ". . . the creation and maintenance of an internal environment in an enterprise where individuals, working together in groups, can perform efficiently and effectively toward the attainment of group goals."

to this approach believe that fundamental management principles are universally applicable to all kinds of organizations and levels of management.

The empirical approach depends upon a study of experience. Case histories are examined to see how their lessons might be applied to future situations.

The human behavior approach concentrates upon interpersonal relations. Management activity is equated with leadership, and the emphasis is upon such matters as motivational techniques.

The social system approach is closely related to the human behavior approach, but emphasizes cultural interrelationships, and theories of cooperation.

The communications center approach views the manager as a sort of switchboard, receiving, processing, and transmitting information.

The decision theory approach sees management primarily as a process of selection from among alternatives, on a basis of rational evaluation. The act of decision-making is viewed as the structural core of management.

The mathematical approach is essentially the same as the decision theory approach, but the emphasis is upon representation of the alternatives in mathematical form. The term "operations research" is often identified with this school of management thought. (Koontz and O'Donnell believe that the mathematical approach should be considered only as a tool, and not as a separate "school.")

Attractions to management careers

Everybody knows that executives make plenty of money. Probably the only question that remains to be answered is the personal one, "How much am *I* likely to make?" It is interesting that the salaries of the top moneymakers are public knowledge.[4] These are the "six-figure men," some of whom make four, five, and six hundred thousand dollars a year. But there are only a handful of such; very few aspiring managers —even ambitious ones—really expect to make the six-figure group. It is more probable that they are thinking of the jobs in the $40,000-plus category, perhaps extending up to $100,000.

According to *The Executive Life*, there seems to be some kind of unwritten formula for determining executive salaries. If the Number 1 man gets $100,000; the next man is likely to be paid $75,000; the third $50,000; and the fourth man from the top, probably $35,000.

---

[4] "Executive Compensation: Who Got Most in '72," *Business Week*, May 5, 1973, pp. 42–44. Every year, *Business Week* publishes the earnings of the top-paid executives in the United State. In 1972, Richard Gerstenberg, chairman of General Motors, was top earner with total earnings of $880,963, including bonuses.

This top-paid group typically will include the president, the executive vice-president, the marketing vice-president, and the financial vice-president. The annual salary of a manufacturing or an engineering vice-president generally would be below this—perhaps at the $25,000-to-$30,000 level. There are no hard and fast rules, and company size is an important deciding factor. In an extremely large company like General Motors, there would be many levels, including half-a-dozen jobs in the $400,000-plus category. However, General Motors can hardly be considered a typical corporation.

Nation's Business magazine regularly surveys trends in executive compensation. In 1972, the mazagine reported that companies with annual sales in the $50 million to $75 million bracket were paying their chief executives $100,000 per year, but as much as $1 billion in annual sales would typically be required before the top man's salary reached $200,000.[5] A somewhat sobering statistic is that most top executives are in their sixties, and so, do not reach the big-money brackets until they are nearly ready for retirement.[6]

While on the subject of pay, it is important to note that the $20,000 to $25,000 bracket has been invaded, in recent years, by people in purely technical categories who have no administrative responsibilities. Obviously, this is partly the result of inflation: $20,000 buys what $15,000 did in 1960. However, there is a more important reason: as industry increases in technical complexity, there is a tendency to give greater recognition to the technical specialist. Usually such a specialist possesses a doctor's degree. A 1972 survey showed that the median salary of engineers with doctor's degrees was about $20,000 per year.[7] It was not too long ago that a salary of $10,000 or $12,000 per year was considered to be in the executive category, but now the median pay of all engineers exceeds $18,000, and a brand-new Ph.D. can *start* at $17,000 or higher.

Management creativity

Many observers have noticed the following trait among management men: after they have earned all the money they can possibly use, they keep right on working as hard as ever. When asked about this, some executives readily admit that salary becomes unimportant above a certain point—that it is only a way of "keeping score."

---

[5] "Executive Pay—Onus on the Bonus," *Nation's Business*, November 1972, pp. 69–73.

[6] When *You* Will Hit Your Peak," *Nation's Business*, January 1963, pp. 64–66.

[7] *Professional Income of Engineers*, 1972 (New York: Engineering Manpower Commission, December 1972), p. 57.

The real motivation of such men must be something other than money and is usually nothing less than a manifestation of an urge toward creativity. A systems engineer has said:

> Before you reach a certain salary level, money is the important thing. After that, job satisfaction takes over.[8]

And from a company president:

> . . . I dearly love this work. You live only one time and you might as well do something you like.[9]

From a labor union vice-president, who talks just like any other executive:

> I'm working harder than I ever have in my life . . . .The incentive isn't monetary gain . . . . I feel I'm part of a crusade, making the world a better place in which to live.[10]

When a management man says his job is creative, he is not just giving idle play to a fashionable word: he means it. He enjoys seeing programs that he originated take shape and prosper, accompanied by organizational flowering and growth. Nor should it be dismissed that many men are motivated by a genuine desire to give service. Almost every human being wishes to feel that his existence has meaning and value to the rest of the world, and managers are no exception.

It is indeed fortunate for the rest of society that the highest reward in management has all the worthy overtones the word "creativity" implies. It certainly provides today's executives with a better motivational creed than the simple one of "profit," which was pursued with such undiluted enthusiasm by the industrial barons of the nineteenth century and led to excesses of greed.

Many executives today—especially those in large companies—believe that corporations have a specific responsibility to the public; they believe that corporate profits must be accompanied by social benefits, and that this must be made a conscious part of company policy. However, there are others who deny that the consciousness is necessary. They believe that only pursuit of the profit objective is necessary and that social benefits will be a natural consequence. In reply, many management men point out that the constrictive government regulations under which businesses must operate today are the direct result of the actions of previous generations of management who believed they had no obliga-

---

[8] W. Guzzardi, Jr., "Man and Corporation," *Fortune*, July 1964, p. 148.

[9] *The Executive Life*, *op. cit.*, p. 69.

[10] William H. Whyte, Jr., *The Organization Man* (Garden City, N.Y.: Doubleday), p. 160.

tions to society other than the pursuit of profit. Some comments on this subject are thought-provoking:

> It is not that they don't care but rather that they tend to assume that the ends of organization and morality coincide, and on such matters as social welfare they give their proxy to the organization.[11]

Louis E. Newman, President, Smithcraft Corporation, has said:

> No greater responsibility do we have than seeing that the skill of managing helps provide a better world than simply reinforce the permanence of power of each manager.[12]

From Thomas J. Watson, Jr., chairman of the board, International Business Machines Corporation:

> Historically, I think we can show that restraints on business have not come into being simply because someone wanted to make life harder for us businessmen. In almost every instance they came about because businessmen had put such emphasis on self-interest that their actions were regarded as objectionable and intolerable by the people and their elected representatives.[13]

From a man identified only as a "thoughtful executive":

> The corporation lives only through the toleration of the people . . . The more estranged the corporation becomes from the majority of people the more likely is the corporation to be the goat when someone wants to make political capital.[14]

Fortune sets up the two opposing ideas in the following way:

1. The main role of management is to do justice between the competing claims of stockholders and other groups.
2. A long-range concern for profits is enough to guide managers.

Fortune then says, "We suggest that the second view is the right one and that it makes the first unnecessary."[15]

---

[11] *The Organization Man, op. cit.*, p. 8.

[12] L. E. Newman, "Managing in a Changing World," *Mechanical Engineering*, April 1964, p. 112.

[13] T. J. Watson, Jr., *A Business and Its Beliefs* (New York: McGraw-Hill, 1963), pp. 90–91.

[14] V. Packard, *The Pyramid Climbers* (New York: Fawcett World Library), pp. 254–255.

[15] "Have Corporations a Higher Duty than Profits?" *Fortune*, August 1960, p. 153.

In spite of *Fortune*'s statement, the prevailing view probably would be that businessmen must strike a balance between the two objectives. *Both* of them bear upon the long-range stability and profitability of the organization. T. J. Watson, Jr. (previously quoted), who is one of the most outspoken proponents of statemanship in corporate management, adds this caution to ensure that the picture remains in balance: "If the businessman fails at business, then all his other concerns will mean nothing, for he will have lost the power to do anything about them."

At General Motors, the nation's largest company, it is fairly well agreed that the concept of social responsibility was not very visible until recent years. But by the early 1970s, after reeling from a series of critical public attacks, G. M. reportedly places social and environmental problems high on its list of priorities. The belief now appears to be prevalent among members of G. M.'s top management that the company ". . . cannot prosper unless it helps to meet the public's social, as well as economic, goals."[16]

The tacit motivations: status and power

One of the most damaging labels that can be fastened onto an aspiring manager is that he is "status-conscious." Yet, virtually everyone is constantly seeking to improve his status, and most human beings enjoy having an influence over their environment; that is, they enjoy the use of power.

The word "status" has acquired undesirable connotations for many people: snobbishness, conceit, egotism, unworthiness, vanity, sham, falsity, pretension, affectation, ostentation. All of these do violence to the basic ideal that all men are created equal. But instead of deliberately choosing unpleasant adjectives to equate to our ideas of status, we could just as well have chosen the words importance, honor, value, esteem, distinction, significance, greatness, quality, respect, and excellence. Expressed in this fashion, a desire for status no longer would seem so ignoble. Putting it yet another way, it can be said that virtually everybody desires recognition; he wants his existence and efforts to be recognized by others as having value. This wish is simply a desire for status, but in more euphemistic dress.

Most people object to the word "status," because they equate it to the activity known as "company politics," wherein advancement through personal accomplishment is abandoned in favor of the more direct techniques of political maneuver, rumor, and insinuation. Yet, it is the *methods* they object to, and not the validity of the goal. Most of man-

---

[16] P. Vanderwicken, "G.M.: The Price of Being 'Responsible'," *Fortune*, January 1972, p. 99ff.

kind has only respect for people of truly outstanding ability, such as great writers, great composers, and—naturally—great engineers and scientists. There are also many industrial managers who have gained public respect, primarily on the basis of their creative accomplishments.

Challenge

After the most basic human wants—food, shelter, and security—have been satisfied, man looks around for new worlds to conquer. If there are no natural obstacles to be overcome, he will invent some. Thus, men compete in business, climb mountains, engage in sports, and write books; many undertake difficult educational programs that go far beyond what would be necessary solely for economic survival.

All these things are manifestations of *challenge*, which, in itself, is one of the most compelling urges that propel men into management careers. It is the excitement and exhilaration of the game itself that some men enjoy. A high level of energy and drive are universally recognized as essential ingredients for management success. A person who does not have this high level of drive, but who aspires to a management career, has already made his first mistake.

Drawbacks in management careers

In any recounting of the "bad things" about management, the following should be recognized: not everyone will agree that these things exist, or even that, if they do exist, they are necessarily bad. Probably, most people who are emotionally equipped to find satisfaction in management would believe these factors to be minor, or perhaps not even relevant to their own cases. A decision concerning the direction of a man's career is, after all, a personal one and will be made according to each person's own value judgments about all the advantages and disadvantages of a given set of alternatives.

LOSS OF PERSONAL FREEDOMS. One infringement on a manager's freedom concerns the right to select social companions as one pleases. Some managers even make it a rule never to socialize with other company people, for fear that such an arrangement might some day prove embarrassing. Conversely, others feel *compelled* to socialize with company people, especially if the company is in a small town. Both conditions are a curtailment of freedom. The greatest casualty is often the executive's wife. Upon her husband's promotion, she is universally advised by management consultants to cut off any friendships she may have made with the wives of men who are now subordinate to her husband. If she finds such action beyond her surgical powers, then she may be subject, later, to vexatious strains generated from the husbands' relationships at work. An equally serious threat is that subconscious compulsions will devolve upon her husband to show favoritism. In any of these events, there has been a curtailment of freedom.

The young executive may find that much of his socializing is in the form of company obligations. For example, he may be required to entertain out-of-town VIPs and may wonder just what all this has to do with his job. Nevertheless, it must be recognized that many executives enjoy this part of their work. While some would regard it as a curtailment of freedom, others would look upon it as a kind of fringe benefit.

Even in such minor matters as personal appearance, there are erosions of freedom. Conformance in dress in itself should not be a serious thing; yet, it seems that a man's chances of advancement are, to some extent, influenced by whether he has short hair, or wears the right kind of ties, or colored shirts, or (continuing to the extreme of trivia) long socks or short socks. For, if minor things like this grate on the boss, they can interfere with a man's promotion. It has been said that "before the boss will promote you, he first has to be able to envision you in his mind's eye as an occupant of the prospective job."

Of a much more serious nature is the demand of practically all companies that an executive give total allegiance to the organization. One student of management behavior has stated, "To get to the top a man must put on a pair of blinders and shut out everything except business. . . . In other words, the corporation must become the life of the man."[17]

While some people would not consider that they had lost anything by fulfilling this demand, others might consider it a loss of freedom.

SUPPRESSION OF EMOTION: STRESS. The executive with ulcers is a standard fixture in the popular image of the modern business world. Like all stereotypes, this one is often false; nevertheless, many managers do experience physical disorders that have their origin in emotional stress. It is pressure, of course, that causes this situation; but what causes the pressure?

In some instances, pressure has been used by the "higher-ups" as a conscious management tool to maintain an atmosphere of urgency and to make sure everyone is working at his maximum output. If a manager objects to the strains on his nervous system, he is likely to be met with the admonition, "If you can't stand the heat, get out of the kitchen." Hence, he is likely to keep his feelings under wraps. The result is more stress.

There are many other well-known sources of stress, such as anxiety concerning job security, or slowness of promotion, or intense compe-

---

[17] Benjamin G. Davis, "Executivism: How to Climb the Executive Ladder," *Mechanical Engineering*, July 1964, pp. 22–25.

tition with rivals, as well as the classic case of the man who is "in over his head" and is struggling to conceal it.

Not so well known, but probably one of the biggest ulcer-producers, is the requirement that the ideal executive always present a calm self-assured façade. Even though, internally, the executive may be as much assailed by feelings of weakness and self-doubt as anyone, he can never allow these to show, or he invites others to trample on him.

Chris Argyris of the Yale Labor and Management Center, who has been very active in research on managers and management characteristics, offers the following as some of the important qualities of the executive:

1. He has high tolerance for frustration.
2. He permits dissection of his ideas without feeling personally threatened.
3. He engages in continual self-examination.
4. He is a strong, cool competitor.
5. He expresses hostility tactfully.
6. He accepts both victories and setbacks with controlled emotions.[18]

After such studious suppression of his emotions as is implied by the characteristics compiled by Dr. Argyris, the executive is then surprised when they flare up in the form of a physical disorder.

Promotion to a position of increased responsibility often brings on a state of mind *Fortune* calls "promotion neurosis," in which the subject experiences great anxiety and emotional conflict. The most common sufferer from this neurosis, says *Fortune*, is the engineer or scientist who has been forced into an administrative job. *Fortune* quotes a psychologist (Harriet Bruce Moore) who says one of the troubles of the engineer-turned-manager is that ". . . he has a very real tendency to regard people (especially his subordinates) as complicated machines which are different from his tools primarily in two ways—they are harder to renovate and more costly to oil."[19]

FAMILY IMPACT. The previously noted demand that the executive put his job before everything else means his family life is often a casualty. In saying this, it must be recognized that not all people assign the same values to the same things. Many men will accept the minimal family life that frequently goes with being an executive and never feel they have missed anything.

One of the things young men very quickly learn is that corporations generally expect instant mobility in their management hopefuls. If a

---

[18] C. Argyris, "Some Characteristics of Successful Executives," *Personnel Journal*, June 1953, pp. 50–55.

[19] *The Executive Life, op. cit.*, p. 87.

man is in Phoenix and his company wants him to go to Omaha (presumably a promotion), it expects him to go without hesitation, and preferably tomorrow, although next week will probably have to do. If he declines, or probably even if he pleads for a delay until June when "the kids" are out of school, then the next man on the list will be chosen, and it can be assumed that this man's climb up the promotion ladder has ceased. Most companies make this clear: mobility is held up as a prime virtue.

The man in question must realize that his wife will have some opinions on this subject. If he has to be in Omaha next week, his *wife* has to stay to look after all the affairs, sell the house, and arrange for moving. If it is deemed important for the children to finish out the year at their present school, she is the one who must stay behind, perhaps even for several months, while her husband makes hurried "commuting" trips home on occasional weekends. Understandably, family relations could become strained under such circumstances, unless both husband *and* wife are thoroughly sold on the same objectives.

A man must examine his (and his wife's) scale of values very closely. As mentioned earlier, up to a certain level (probably through the "superintendent level" already described earlier in this chapter and, in rare cases, even into the "manager level") a man can "have his cake and eat it too." If he has advancement aspirations beyond this, but insists that his family come before the company, he will almost inevitably come to the point where he will be forced to choose.

If a family is one in which both husband and wife are professionals, matters become infinitely more complex. Each may have career aspirations which are not likely to develop neatly in a coordinated way. Yet, surprising accommodations can sometimes be arranged, such as the one reported by Alva Matthews, who is a consultant for an engineering firm in New York City. She works three days a week, living in the city of her husband's job, Rochester, and commuting regularly to New York for her job.[20]

POLITICS AND JUNGLE-FIGHTING. Politics and "jungle-fighting" do exist, although not to the extent suggested by popular fiction. Since they are probably the most thoroughly publicized and well understood of management drawbacks, not much time will be spent on them here. Nothing can protect a man from falling into the mistake of using these fabled evils, except a strong sense of personal integrity. If a man should discover that he is spending more of his business day with actions that have "getting ahead" as their object, rather than doing the job well, he is in danger of taking the next step into the arena of political maneuver.

---

[20] J. Kotel, "The Ms. Factor in ASME," *Mechanical Engineering*, July 1973, p. 16.

Among the commonest of such maneuvers is the skillful discrediting of an opponent by any one of many techniques, such as planting rumor, sowing doubts, withholding information, maneuvering him into an untenable position, and methods of a similar nature. An equally common maneuver is the seeking to make one's self more apparent; that is, to attract favorable attention from one's superiors. The techniques are almost as numerous as are their practitioners. They include such procedures as marking the boss's *boss* for copies of memos, finding excuses to visit headquarters (in the case of a branch office man), currying favor with the boss, or being especially agreeable to the boss's secretary.

There are even coaching services available to executives who feel deficient in political skills. Vance Packard reports on one firm that instructs its clients, among other things, to keep a file of index cards on "important people" and to edge out into social spheres. This firm goes so far as to suggest, "Follow your immediate vice-president into your favorite bar and have a drink with him." Packard flatly gives his opinion that if the man dutifully follows these precepts, ". . . the coaches will have succeeded in producing a real grade-A corporate creep."

A CONFLICT OF MORALITIES. A serious potential difficulty in a management career arises in the sphere of moral action. The problem comes into being because many companies demand that their managers follow a rule of seeking only the good of the company, to the exclusion of other considerations. A three-year study, conducted at the University of California, Los Angeles, on executives and how they get ahead showed some executives were not sure that the usual moral standards observed by most people in their personal lives are applicable to business.[21]

That such a policy often backfires, has already been mentioned. When the policy expresses itself in the form of trusts and cartels, society reacts by passing antitrust laws. When the policy results in the exploitation of working men, labor reacts by forming unions. When it expresses itself in the form of cheap, unsafe equipment, government agencies step in with restrictive laws.

In the price-fixing suits against the electrical industry in 1961, the presiding judge described the individual defendants as "torn between conscience and an approved corporate policy . . . the company man, the conformist, who goes along with his superiors and finds balm for his conscience in additional comforts and the security of his place in the corporate setup." It was shown that, within the organizations, managers who believed in obeying the law were sidetracked from promotions; their respect for ethical behavior was mocked as "religious." One of the defendants stated: "I was to replace a man who took a

[21] R. M. Powell, "How Men Get Ahead," *Nation's Business*, March 1964, p. 58.

strictly religious view of it; who, because he had signed this slip of paper wouldn't contact competitors or talk to them—even when they came to his home." He added, "I was glad to get the promotion. I had no objections."

Fortune commented on these suits: "No thoughtful person could have left that courtroom untroubled by the problems of corporate power and corporate ethics. . . . Big business . . . establishes the kind of competition that is typical of our system and sets the moral tone of the market place."[22]

It should not be inferred that a man must discard his moral code to succeed in management. Encouragement can be drawn from the fact that some of America's most successful managers are also well known for their adherence to high sets of standards.

Getting there

Compare the following statements:

> Forty percent of management is recruited from the ranks of the engineers.[23]

> Some time ago I noticed that appointments to top-level positions were seldom going to engineers, but were going to finance men, marketing men, and lawyers.[24]

> The proportion of top industrial management with a background in science and engineering will have risen to more than 50 percent by 1980. U. S. industry is indeed coming under new management.[25]

> Despite the swift advances of technology, there has apparently been a drop-off in the proportion of industrial scientists and engineers who reach the position of chief executive officer.[26]

Strangely enough, *all* of these apparently contradictory statements are true. As might be expected, the paradox results from the interpretation. A study of top executives made by *Fortune* in 1952 showed that only 10 percent of them came up through engineering; the same study showed that 46 percent of them had studied science or engineer-

---

[22] R. A. Smith, "The Incredible Electrical Conspiracy," *Fortune*, April 1961, p. 132ff.

[23] H. W. Dougherty, *Your Approach to Professionalism* (New York: Engineers' Council for Professional Development, 1959), p. 43.

[24] Davis, "Executivism," *op. cit.*, p. 22.

[25] *U. S. Industry: Under New Management, A Scientific American Study* (New York: Scientific American, 1963), p. 33.

[26] *The Pyramid Climbers*, *op. cit.*, p. 170.

Table 7–1 Executive backgrounds

	Percent	*Totals, percent*
Science or engineering		
Bachelor's degree	26	
Graduate degree	6	
No degree	5	
Nonscience		37
Business degree	6	
LL.B. degree	12	
Other college degree	23	
No degree	22	
		63
		100

Source: *U. S. Industry: Under New Management.* (New York: Scientific American, 1964), p. 31.

ing in college.[27] Yet, 25 percent of the top executives had risen through sales, 23 percent through production, 17 percent through finance, 16 percent through general management, and 8 percent through law. What accounts for the apparent contradictions in the statements at the beginning of this section, is an ambiguity in the use of the word "engineers." In one case it means those who are actually working as engineers in the company, while in the other, it means all those who may once have studied engineering. The implication is that it is fine to take engineering while in college, if you yearn for the corporation president's job; but you should not have very high hopes of getting there through the engineering department.

A *Scientific American* study made in 1963 gave the results in Table 7-1. (The study included data on 800 executives, selected as the top two men in each American firm having annual gross sales of $100 million or more.)

Scientific American's studies show that the percentage of top corporation officials with a background in science or engineering has risen from 20 percent in 1950 to the 1963 level of 37 percent. From the composition of the pool from which future management will be drawn, *Scientific American* predicts that this percentage will increase to more than 50 percent by 1980.

The view has occasionally been advanced that there is something

[27] *The Executive Life, op. cit.*, pp. 31–34.

about an engineering education which gives one a special advantage in a managerial career; further, the inference is drawn that engineers hold such an advantage because they can apply the "scientific method" to their management activities. But there is dispute over just what the "scientific method" is, and at any rate it is hard to see how it applies to management. Yet, the results of the *Scientific American* study cannot be denied. What does it all mean?

Part of the explanation is that science and engineering are difficult courses of study, and successful completion of a college program in one of these curriculums means that the student has learned to work hard and successfully. He has already developed, to a high level of proficiency, his personal initiative and his ability to finish the jobs he has undertaken. These statements could apply equally well to a law degree or to a master's degree in business administration; however, these two educations take longer than four years and, thus, attract fewer people.

What really gives science and engineering the edge is the growing technical complexity of industry. This statement should not be misunderstood: it does *not* mean that the top-level manager must have a large fund of detailed technical know-how. It means that the top man in a technically based industry needs a technical background as part of his *cultural* understanding. The word "cultural" is used here in the broadest possible way: it refers to all the knowledge and experience a man possesses, against which he compares his current situation in order to make value judgments. Many industries today completely depend upon rapid technological change for their well-being; it seems inevitable that managers of such enterprises will possess considerable technical understanding, and that this trend will increase.

If an engineering education is at least as good as any other preparation for a man who wants to be a company president, then what comes next? Finance and law would appear to be closed routes for the engineering graduate, although they are classic routes to the top. Still, *Fortune*'s 1952 study indicates that sales and production are even better routes than finance and law; almost half of the executives studied by *Fortune*[28] had achieved their positions through the sales and production departments. These two routes are wide open to the engineer, and in fact, many new engineering graduates go directly into one of these fields.

Many young men ask if it would be wise to get a Master of Business Administration degree if they are interested in management. Insofar as a general answer to such a question is possible, this would probably be it: if a man feels he is destined for top management, an M.B.A. would be extremely useful to him. However, many companies seem to believe that a B.S. is just the ticket for management aspirants. In any event,

---

[28] *The Executive Life*, *op. cit.*, pp. 31–34.

it should be remembered that an individual's personal characteristics and drive are more important than mere possession of degrees.

There is a small financial advantage in starting salaries for those with the M.B.A., as compared to those with a master's degree in engineering, but hardly enough to constitute an attraction in itself. In 1973, the average starting salary for a person with an M.B.A. plus an undergraduate degree in a technical field was $1177 per month. The average for engineering master's degrees, depending upon field, ranged from $1020 to $1093.

The recession and business readjustments of 1969–1971 apparently removed some of the former allure of the M.B.A., at least for some companies. In 1972, a recruiter for a manufacturing company declared, "Four or five years ago an engineer with an M.B.A. was a glamor boy. Now we prefer a man with just an engineering degree."[29] In spite of such a pronouncement, only one year later the nation's leading graduate business schools reported a bull market for their M.B.A. graduates. Schools such as Harvard, Stanford, and Wharton said their graduates were going to positions paying an average of $1400 per month. Most of them went to banks and accounting firms, although some went to large manufacturing firms like General Motors, Hewlett-Packard, and Xerox.[30]

Another kind of graduate program is accessible to engineers with management aspirations. These are the master's programs in engineering management, offered by a dozen or so universities in the United States. (Some also offer bachelor's programs.) The prospective master's student in engineering management is usually expected to have a B.S. degree in engineering, and studies a combination of management and technical courses. Typical courses are organizational theory, marketing management, production management, human relations, accounting, engineering law, project management, value analysis, computer science, and operations research. Most often, graduates from these programs obtain their first jobs in production, industrial engineering, or marketing.[31]

What it takes

Many of the essential managerial characteristics have already been discussed. To summarize, these have been:

1. A willingness to place the company first
2. A high degree of aggressiveness and drive, including a willingness to work long hours

---

[29] "Too Much Learning . . .," *Forbes*, July 15, 1972, p. 42.

[30] *Time*, August 6, 1973, p. 65.

[31] D. L. Babcock, "B.S. and M.S. Programs in Engineering Management," and D. B. Smith, "Graduate Engineering Management With Flexible Options," *Engineering Education*, November 1973, pp. 101–104, 108ff.

3. An ability to handle others
4. A strong desire for personal status and economic gain
5. A desire to be in control
6. A high degree of tolerance for frustration and disappointment

To these should be added the following:

7. Persistent optimism. (No matter how bleak things look, the ideal management man always has a constructive program on tap for which he entertains the highest hopes. The cynic is unpopular in management circles.)
8. The ability to *finish* a job, as well as to initiate it. (Actually, this characteristic is in demand at all times and places, and not just for management positions. It is among the rarest of the world's commodities, and its absence is seldom detected by people who do not have it.)
9. Good judgment and logical thinking ability. (These qualities are sometimes known simply as "intelligence." In a complicated situation, the manager must be able to sort things out into their proper relationships and to dig beneath surface irrelevancies to get at the heart of issues. This having been accomplished, he must forecast the future and be right most of the time.)
10. The ability to communicate, not only in writing but above all, orally.

Vance Packard offers these four basic "survival" rules for managers:[32]

RULE ONE:	Be Dedicated
RULE TWO:	Be Loyal
RULE THREE:	Be Adaptable
RULE FOUR:	Be Quietly Deferential

Concerning Rule Four, it probably should be added that, while nobody likes a "yes-man," nobody likes a "no-man," either. This is a ticklish matter and can cause a man a certain amount of ethical queasiness: In essence, a man must appear, to the boss, to be on the boss's team most of the time and use his own creative input with discrimination.

A prominent question in management is how much importance should be attached to technical know-how. In recent years, there has been a tendency to disparage the need for subject-matter competence on the part of a manager. Instead, it is claimed that the manager should

[32] *The Pyramid Climbers, op. cit.*, pp. 103–110.

be an expert at "management skills"; presumably, these include proficiency in such things as human relations, budgeting, and planning. When applied to a *general* management job, the advice makes considerable sense; when applied to the manager of one of the functional branches of the company, the advice seems defective.

A *general* manager, who must simultaneously manage several functional branches (for instance, sales, finance, manufacturing, and engineering), will find it a physical impossibility to be expert in all of the departments he is managing. If he should persist in acting as an expert in the one he does know (the branch through which he came up), then he is a thorn in the side of everyone who must work under him in that branch. Obviously, he needs *some* background in all branches (gained partly through college course-work, but more importantly through experience); however, he cannot be an expert in all. Here the advice at issue is good.

Many promoters of this advice insist that it should apply to *all* levels of management, but the reasoning applied here to general management jobs is not necessarily applicable to managers of the functional branches, such as general sales manager, director of manufacturing, or chief engineer. In these cases, it is preferable to have a manager who is technically competent in his own area, as well as in possession of the requisite management abilities. Of course, he must be capable of suppressing the urge to function in a technical operating capacity himself. He must get others to do the actual operations, but he will be a better manager if he himself thoroughly understands those operations.

Advice of the opposite extreme, which holds that the *best* technical man should be chosen leader, is equally bad. This point of view is usually espoused by someone who thinks *himself* to be the best technical man. The notorious failures of this approach have undoubtedly helped to form the opposite attitude. The following procedure would seem to make the most sense (at least until the level below that of general manager): first, select a group of men for their thorough competence in the subject-matter to be managed; then, from these, make a final selection on the basis of their managerial skills.

One last matter is the question: should a man join one company and stick with it, or should he shift around a little during his early working years? Nobody knows for sure. The *Fortune* study on executives[33] revealed that a full third of the subjects had never worked for another company and an additional 27 percent had worked for only one other company. Almost half of the subjects had been with their current employers more than 30 years. The least mobility was shown by those in the oil industry. The most mobile were steel executives, with automobile executives close behind them.

The biggest single danger in staying with one organization is that a person could find himself becoming more and more highly specialized in one particular thing and, correspondingly, more and more narrow.

---

[33] *The Executive Life, op. cit.*, pp. 31–34.

Most organizations will encourage such specialization as long as it is useful but will be quick to amputate it when it is no longer needed. This practice is the main worry that causes young men to move during their early years, and many companies combat it by offering unusually high salaries for the specialties currently in demand.

Some large companies are in a position to avoid these disadvantages. They can transfer their men frequently among divisions and, thus, give them the same broad experience the men would acquire if they were to work for many different employers. This is great for the management-bound man, provided he is "geographically flexible." On the other hand, a capable young man can often shine to better advantage if he is in a small company, especially if the company is a fast-growing one.

Engineering is *management*

There is sufficient similarity to give substance to the assertion that engineering is part of management. Consider the president of a corporation: in his daily activities, he must conserve money, conserve time, maximize effort, increase the competitiveness of his company's products, and, finally, satisfy the conflicting internal requirements of the company's various departments, such as sales, service, manufacturing, finance, and engineering.

The preceding is also a fairly accurate enumeration of the opposing forces the engineer must balance in fulfilling *his* responsibilities. The president has additional responsibilities, of course. No one would seriously suggest that the engineering department duplicates the president's function, but there is an especially important characteristic belonging to the president of the corporation which is worth noting: he is a man who *cares*. In this crucial aspect, the engineer's outlook should exactly match the president's. He must care deeply about what is going to happen to the organization. Without any exaggeration, it can be said that the future of the company has been laid in the hands of its engineers, for the future consists of the new products the engineers will create. It is entirely possible for the engineering department to make or break a company.

Is it the conception of spectacular new inventions that has so great an influence? Although such conception is essential, it is by no means the whole story. A program may begin with a concept for a great new product, but there are numerous places along the development route where it can be ruined. Probably the most sensitive movement in the entire life-history of a potential new product occurs during the selection of the basic physical method by means of which the functional objectives will be achieved. Two different products with similar functional purposes can differ enormously in their ultimate success, depending upon the physical make-up of their internal systems.

No doubt it will be argued that higher management will have the ultimate word on these matters and that the only function the engineering department can have is to make proposals. This is true, but higher management can operate only on the proposals submitted. They can refuse bad proposals, but the act of refusing does not create good ones. If the engineers are unable to create good proposals, there will be no new products—unless higher management is willing to accept the least objectionable of the poor proposals, perhaps never even knowing these are poor proposals, since they will not have seen any better ones.

After several possible systems have been proposed, another critical phase is approached. It is at this point that the engineer's knowledge of mathematics and science comes into play. Somehow, each system has to be reduced to an estimated cost, since it is upon *cost* that higher management's decisions will be based and it is also upon *cost* that the ultimate competitive vitality of the product will depend. To get this information, each system must be analyzed on paper before anything is actually built. Later, limited experiments may be conducted to settle some fine points, but at this stage, everything depends upon what goes on in the engineers' heads. What are the most influential parameters in the system, and how might one select a reasonable criterion for optimizing them? How many components, and of what quality, will be necessary to accomplish each system? What will be the influence upon performance (and cost) if the component characteristics vary from their nominal values? What about strength, wear, corrosion? How much additional (and expensive) quality is justifiable in the initial manufacture of the produce so as to avoid undue field service costs later?

Nobody can duplicate the engineers in this role, unless he himself sits down and repeats the whole process, step by step. If the engineers do their job wrong it will not be of much help that everyone else has done his right. Moreover, nobody will know if the engineer's job has been done right or wrong until it is too late.

A key person: the project engineer

In any given engineering department, it is usually easy to determine who the key people are. It should not be supposed that the key group consists exclusively of supervisors. Many key people will turn out to be supervisors, of course, but not all. However, all the key people are distinguished by two important characteristics: first, they will be outstandingly competent at the *technical* aspects of their work; and second, they will be men of such stature and authority in that they are the ones to whom people naturally turn for direction. For the most part, they will be first-line engineering supervisors, a category frequently referred to as "project engineer." (However, it should be noted that this particular title has widely varying meanings in different companies.)

Project engineers make the real engineering decisions. Engineers at a *higher* administrative level are usually too remote from daily activities to be able to bring sufficient detailed technical understanding to the decision-making process. Engineers at a *lower* level lack the authority and overall knowledge to be able to make decisions that will stand up in the long run.

Key men like these fill the pivotal posts of engineering throughout the United States. Collectively, their actions have an almost incalculable effect upon the rest of the country, since they possess a combination of technical knowledge and authority which is sufficient to control the directions that engineering technology will take and, hence, to determine the nature of the goods that will be made available to society.

It would be a strong temptation simply to classify these men as managers, for they usually do have others working for them—as few as three or four or as many as 15 or 20. But it is essential to note that at least 50 percent of the project engineers' time is spent in directly technical affairs. This is what makes them influential: they are fundamentally *creators*.

Surveys show that most engineers wind up as managers of one sort or another. The National Engineers Register for 1969 shows the following breakdowns of responsibility as a function of age:

		Age	
	30–35	*40–45*	*50–55*
General management	4 percent	12 percent	15 percent
Manager of major divison	12	24	27
Project supervisor	25	24	20
Unit supervisor	16	12	9
Indirect supervisor	21	16	16
No supervisory responsibility	22	12	13
	100	100	100

Thus, by age 35, more than half the engineers surveyed could be considered managers, and by age 45 over two thirds of them could be so classed.[34]

[34] *The Engineer as a Manager* (New York: Engineering Manpower Commission, September 1973).

EIGHT

Organizational relationships

Profits (and losses)

Some corporate officers have been known to state that their only business principle is "to make a profit." Such a policy does correctly describe the justification for a corporation's existence. It is for just such a purpose that the stockholders put their money into the business.

In the larger socioeconomic sense, a corporation's purpose is to supply the goods needed by people. In Russia, the entire economy is run like a single gigantic company, with managerial directives and a system of merit awards relied upon to achieve efficiency. In the United States, it is believed that the economic system works better when it is based upon privately owned competitive business units, coupled with the incentive of profits. To keep the greed for profits from overbalancing the benefits to society, Americans attempt to regulate the system by means of taxation and regulatory laws.

In Chapter 7 it was mentioned that a preoccupation with short-range profits can bring a company to grief. Short-range profits are important, of course; without them a company will find it unnecessary to worry about long-range profits because it will soon have ceased to exist. But corporations are presumably in business on a long-range basis, and for this reason, the managers worry about the status of the corporation as a good "citizen." They wish the company to be well thought of for a variety of reasons: so that it will inspire the confidence of customers; so that it will attract, and hold, good employees; and so that it will avoid the unfavorable attention of legislative bodies. However, the managers of a corporation are entiled to worry about such things only because they believe this is the best way to ensure the long-term profitability of their organization.

Viewed in this light, engineering must be regarded as a very heavy expense item that is justifiable only if it improves chances for future profit. Yet in recent years, virtually all manufacturing companies have been energetic in increasing this expense item, because they know they cannot have new products without engineering—and they know, further, that their future depends upon new products. In 1964, Booz, Allen, and Hamilton stated that many companies were finding that more than half their sales were coming from products which had been unknown 10 years previously.[1] Furthermore, on the basis of their wide management consulting experience, Booz, Allen, and Hamilton asserts that, on the average, only one idea out of every 58 evaluated actually becomes a successful product. Even worse, they state that, out of every three products that are "successful" at the Research and Development (R & D) stage, there emerges only one commercial success. Obviously, then, it becomes of overriding importance to improve the initial product selection processes to the highest degree possible, *before* projects get into the costly R & D stage. This matter will be treated later, in the section on developing new products.

Internal relationships

Occasionally, people have been known to ask, "Which is the most important branch of a company: sales, engineering, or manufacturing?" One might as well ask, "Which is the most important leg of a three-legged stool?" Obviously, all are essential. Briefly, each group's special functional contributions to the organization are as follows:

Engineering *defines* the product.
Manufacturing *makes* the product.
Sales (or marketing) *distributes* the product and is the income-producing agency of the company.

These distinctions may seem obvious, and even simple-minded, but it is precisely because many people overlook such fundamentals that strife sometimes occurs among departments. Two examples will suffice to give an idea of the manner in which problems can arise and the ways in which they may be avoided:

EXAMPLE 1. Engineering releases a set of drawings for a new product. After a lapse of time, Manufacturing reports that Engineering's design cannot be produced. The engineering department, which has anticipated this, whips out a prepared method by which production can be accomplished. Manufacturing, after due examination, declares the method impractical. Engineering thereupon announces that it will

---

[1] *Management of New Products* (New York: Booz, Allen, and Hamilton, 1964). Booz, Allen, and Hamilton are a prominent engineering consultant service.

establish a model department of the proposed process and prove its practicality. If it turns out to be successful in this endeavor, Engineering now finds itself in the production business and is faced with two alternatives: 1) it can remain permanently in the production game or 2) it can somehow convince Manufacturing to assume control of the "orphan" department. Either way, relations between the two departments are apt to be less than cordial. Earlier consultation and cooperation with the manufacturing department would have avoided most of the trouble.

EXAMPLE 2. On a new design, Manufacturing suggests some changes and claims it will reduce the cost of the product. After examination, Engineering declares the original design is better and rejects the proposed changes. Manufacturing, without authorization, makes the changes anyway, The changes prove successful, and Manufacturing continues to produce the article the new way and says nothing to Engineering. Manufacturing is now in the engineering business, but nobody else knows it. If this situation occurs very often, confusion and costly errors lie ahead, plus an inevitable blowup and deterioration of relations between the two departments.

The prescription for improvement is simple; it is based upon only two principles: 1) each department should bear constantly in mind what its basic function is and adhere to it; 2) each department should respect the special competence of the other and seek the consultation of the other on a continuing basis.

Some attempts have been made to circumvent these problems by placing a small engineering group *within* a manufacturing department, with authority to make changes on products. However, this usually changes nothing. The new engineering group soon develops its own sense of identity, and the same kinds of problems can take place as before, except this time they occur between the new group and the rest of Manufacturing. The prescription for improvement is the same as that previously given.

The greater burden for preserving effective relations rests upon Engineering. The very nature of their relationship places manufacturing departments in the position of continually receiving "orders" from Engineering, in the form of drawings and specifications. Furthermore, most engineers have college degrees, whereas most manufacturing men do not, even though they may possess the native ability to succeed in college had they gone there. For these reasons, manufacturing personnel automatically are likely to feel at a disadvantage. The slightest indication of superiority on the part of engineering personnel may be enough to tip the scale in an unfavorable direction. Almost every manufacturing organization can cite examples of engineers who are unable to work with manufacturing personnel, usually because of an unconscious (sometimes *conscious*) attitude of superiority which they convey.

A senior student in engineering at the University of California, Davis, chose the subject of engineer-machinist relations for an oral report he was to present to a group of classmates. To prepare for his talk, this student solicited responses from 20 machinists working at various experimental shops in northern California. He was hardly prepared for the emotionally charged responses he received. Following are representative answers to his question, "What, in your opinion, is the typical engineer's attitude toward machinists?"

> That we are dirt under his feet—that the machinist doesn't know anything.

> Lacks respect for machinists. That machinists are inferior.

> In too many cases the actual design is done by the machinist and the engineer takes the credit for it.

On the other hand, virtually all the respondents expressed high respect for engineers who act as though machinists have something of value to contribute, but implied that engineers of this type are too rare.

Engineering's relationships with Marketing tend to be of a more subtle variety than are those it maintains with Manufacturing. Here, if any conflict arises, it is likely to be over product function and customer acceptance. For example, Marketing is apt to believe it has a better appreciation of what the customer wants in a product than does Engineering, while the engineering departments may think the customer's demands are illogical. Having been encouraged all his life to think logically, an engineer may believe it only reasonable that customers should do the same. Customers, of course, remain unimpressed; they simply buy a competitive product if they feel so inclined.

The matters in question may be things like appearance, color, noise, arrangement of controls, and the method of use. These items definitely come under the heading "defining the product," which is Engineering's basic responsibility. No clear-cut boundary is possible, yet marketing people generally believe that their department should be the principal authority on features such as those just mentioned, all of which profoundly affect marketability. To justify this view, one marketing vice-president of a national company was fond of relating a story whenever his engineer associates were proving to be contrary in yielding a point. His anecdote concerned a dog food company that had invested a considerable sum of money in developing the perfect dog food. After years of effort, the company's scientists at last proclaimed success. Their new product consisted of a carefully balanced formula of proteins, fats, and carbohydrates and contained all the essential vitamins and minerals. Close attention had been given to the perfect blending of many savory aromas and the achievement of an attractive texture. Furthermore, it was economical to manufacture. "The only trouble," said the marketing vice-president, "is that the dogs wouldn't eat it."

Line and staff

In almost any organization, questions will arise concerning which groups represent "line" and which "staff." Supposedly, the "line" groups are those which carry out the basic business activity of the company; all other groups support and/or advise those activities, and are thus considered "staff." Students of organizational structures might have no difficulty in dividing a company into line and staff groups, but the members within those groups often tend to dispute their classifications. It would seem that most people want to be considered as in line groups, for along with that designation goes an implication of indispensability. Many people fear the classification as staff, believing that someone might someday decide their particular support function is no longer essential.

In a manufacturing organization, the production departments are obviously line functions. But how is the purchasing department to be classified? Is it a support function for the production departments (hence, staff), or is it the first and indispensable group in initiating production (hence, line)? It could be considered to be either. As for the marketing department, most people would consider it to be a line function, but what is the engineering department? No doubt most students of the subject would consider engineering to be a staff function, but many engineers become irritated with this and point out that there would be no manufacturing or marketing divisions at all if it were not for the products designed by the engineering department.

The reader might ask, in all this excitement concerning line and staff, just what difference does it really make? Some have answered: not much. But others disagree and claim that a great deal of organizational friction occurs because of failure to distinguish successfully between line and staff activities. They point to the common problem of a staff adviser who oversteps his bounds, and attempts to issue instructions to line managers. Occasionally, staff officers have justification for exercising a certain amount of control over line managers, as when a personnel manager requires that a supervisor adhere to prescribed seniority policies, or when an industrial engineer requires that a shop foreman follow established manufacturing procedures. But these are classic sources of friction, and it is a wise manager who knows how to maneuver through the maze of overlapping line and staff relationships.[2]

---

[2] H. Koontz and C. O'Donnell, *Principles of Management: An Analysis of Managerial Functions*, 5th ed. (New York: McGraw-Hill, 1972), pp. 302–324.

Developing new products

This section will deal with one of the most difficult problems facing modern industry: how one goes about developing successful new products. Today it is well known that any successful product has only a limited life. Booz, Allen, and Hamilton claim that when a product reaches its peak in terms of sales volume, it has already started its downhill slide in terms of profits.[3] In the current world, a continuing stream of new products is absolutely essential to the well-being of a business.

Good sense indicates that every new product proposal should be analyzed in terms of its prospective return on investment. Those products with the greatest apparent return and the lowest apparent risk are selected for development. Although fine in theory, this precept requires prediction of the future, and just how does one predict the future?

Corporations' usual answer to the preceding question is to carry out a survey. But market surveys are generally successful only in ascertaining the *current* state of affairs and then, only moderately so: often, they are useless at predicting what the customer will do in the future—the customer himself does not know what he will do. If the proposed product under survey is *really* new, so that nobody has ever heard of it before, then a market survey will generally produce nothing but a vacuum—precisely because nobody ever heard of the product before. If a sizeable market is forecast, it at best represents someone's guess, however intelligent and informed such a guess may be. Yet, how can one calculate a return on his investment (which is a reasonably scientific process) when he can't get a scientific grip on the size of his market? If there is a shortage of capital, which is often the case, the pressure may be all but irresistible to put available funds into the area of greatest certainty; this probably means following a trail someone else has blazed; and if the organization yields, it has then chosen to be a follower instead of a leader, however unwilling its decision may have been.

Top management usually takes the viewpoint that one of its most important functions is to foster, analyze, and develop new product ideas. Hence, there is a tendency to keep the new product function very close to the top, organizationally. If the new products in question are mostly extensions of the present product line, these responsibilities may be entrusted to the marketing vice-president. However, if the new product investigations reach into areas that are completely unexplored, such responsibilities more likely will be given to a new products department, sometimes directly attached to the president's office.

Generally, the functions of a new products department will be to:

---

[3] *Management of New Products, op. cit.*, p. 4.

1. Generate new ideas
2. Analyze and screen ideas from all sources
3. Coordinate market needs with research and development ideas
4. Conduct pilot tests of new products, under market conditions

Of these four functions, the most important is probably the second. Most often, companies rely upon their research and engineering departments to generate new ideas (function 1).[4] Sometimes new ideas come from outside the company, but nearly all observers agree this is a minor source. Other possible sources of new ideas are:

1. UNUSED PATENTS. A veritable avalanche of new patents is issued every year; very few of these patents see commercial use because, for numerous reasons, patent owners may find themselves unable to press ahead with commercial development. As a result, many patents are available for license or sale. In addition, an examination of issued patents may generate entirely new product ideas. If one wished to find a route to the largest number of creative minds in this country, the trail of issued patents would surely be his best starting-place. The Small Business Administration, Washington, D.C., regularly publishes a *Products List Circular* containing abstracts of patented inventions available for sale or license.

2. GOVERNMENT NEEDS. Supplying goods needed by the government is big business. The gigantic aerospace industry largely depends upon the government as its principal customer. Major aerospace development companies have special Washington offices, whose purpose it is to maintain constant contact with government agencies in order to ascertain their needs and to anticipate trends. Each year, the government issues a publication, *Inventions Wanted by the Armed Forces*, which can be an important stimulant to new product ideas.

3. TOP MANAGEMENT. In some organizations, particularly those which were founded by creative technical people, a major source of new ideas continues to be the top man or men.[5]

Whatever their source, after a group of ideas has been collected, they must be carefully screened (function 2). Out of this group, an average of only one in 58 actually becomes a successful product, according to Booz, Allen, and Hamilton.

---

[4] In the American Management Association's book, *Developing a Product Strategy*, 34 executives from 29 different organizations have presented their views on the new products function. This book is an important source of the information in this chapter.

[5] W. Wade, "Supplementary Opportunities for Technical Innovations," in *Developing a Product Strategy, op. cit.*, pp. 125–130.

Ideas that survive the screening step are analyzed on a return-on-investment basis. At this point, an estimate of the future market *must* be made; usually this estimate is coupled with a market survey.

As was explained earlier in this section, the information acquired through a market survey must be examined with an experienced and skeptical eye. Many companies have been seriously misled by accepting survey results at face value. One incident, reported by a vice-president of Westinghouse, concerns portable television receivers. When portable receivers were first being considered, two electronics manufacturers conducted surveys on the receivers' market potential and both received negative reports. One company abandoned the idea, the other went ahead. Today, of course, portable television sets are a major product.[6] The point of the story is that the wants of today (which are mostly what a market survey will reveal) may be quite different from the wants of *tomorrow*.

It should not be inferred from the foregoing that market surveys are useless and a waste of time. On the contrary, survey information has many times proved extremely useful when viewed in a careful and discriminating light. Almost invariably, unexpected—and sometimes startling—information will be turned up.

In its *coordinating* function (function 3), the new products department is expected to take an overall view of the company's operations and to provide a balance-wheel for any excessive enthusiasms generated by salesmen and by engineers.

The general criticism against engineers is that they are overenthusiastic about the technical improvements their ideas offer and insufficiently responsive to what the market wants and is willing to pay for.

The criticism against most salesmen is that, while they automatically wax enthusiastic over *any* new idea (a generally desirable quality in a salesman), this enthusiasm blocks true objectivity. Another point against relying upon line salesmen for new product ideas is that they tend to think too much in terms of the sale they could have made last week if they had had such-and-such a feature and do not think broadly enough in terms of the future.

Many products fail because of insufficient field testing under actual market conditions (function 4). The customer can be relied upon to find all sorts of defects in the product that have somehow not shown up during in-plant testing. More efficiently conducted, field testing can show whether the wants of the market place have been correctly identified, *before* costly and nearly irrevocable steps have been taken toward mass production and distribution.

An example which illustrates the point just mentioned, concerns a special bottle-warmer that Westinghouse was planning to place on the

---

[6] C. J. Whittling, "Gauging the Potential of a Product or Product Change," in *Developing a Product Strategy, op. cit.*, p. 134

market. The warmer employed an unusual principle (thermoelectrics), and included a timer that would cause the heating cycle to start at a predetermined time, so that the bottle would be warm when the baby's parent got up for the 2 A.M. feeding. Westinghouse undertook extensive tests, interviewed department story buyers and sales people, and conducted panel discussions of young mothers. Finally, after carrying out actual field tests in various areas at different price levels, Westinghouse shelved the product. From the field tests, it was found that the product could command only a price in the $8-to-$25 range, whereas the company planning staff had envisioned a luxury $30-to-$50 range.[7]

Return on investment analysis

Before a project goes very far, someone must take a good hard look at the future and try to decide if the proposed product has any chance of producing a profit. In order to do this, two things are necessary: 1) somebody must devise a reasonably clear configuration of the proposed product, so that a production cost can be estimated; 2) somebody must predict the anticipated annual sales.

Table 8-1 shows a typical return-on-investment analysis for a consumer product that will sell for $300. For this product, it is estimated that a million dollars will be required for development and another million for plant expansion, tooling, and distribution start-up (including sales training costs, service training costs, and the expense of staffing new offices). Such a capital investment is by no means unusual; in fact, the usual tendency is to underestimate the amount of capital required.

In the analysis shown, it is assumed that the third year after introduction of the product represents what will be a steady-state condition and that the return of 27 precent per year will continue henceforth. At this rate, the company's entire investment would be recovered by the fifth year. However, such an assumption might be naïve. For one thing, the competition cannot be expected to stand still during this five-year period, and there is a high likelihood that additional R & D investments will be required during the five years to keep ahead of the competition. Viewed in this light, even a 27-percent annual return could be regarded as a borderline profit, and the company perhaps should look for other products that might afford greater returns and faster recovey of capital. Furthermore, if the sales estimates are optimistic, as often seems to be the case, or the manufacturing cost estimate turns out to be too low, the whole proposition could become quite unattractive.

---

[7] C. J. Whittling, *op. cit.*, pp. 137–138.

Table 8.1 Sample of a return-on-investment analysis for a consumer product

	First year	*Second year*	*Third year*
Initial investment (R&D, plant expansion, tooling, and so forth)	$2,000,000		
Number of units required for sales inventory	4000	8000	12,000
Dollars tied up in inventory, at factory value	($150 × 4000) = $600,000	($100 × 8000) = $800,000	($90 × 12,000) = $1,080,000
Total capital investment	$2,600,000	$2,800,000	$3,080,000
Annual sales (number of units)	3600	7200	12,000
Manufacturing cost, per unit	$150	$100	$90
Selling price, per unit	$300	$300	$300
Selling expense (at 37½ percent)	$112	$112	$112
General and Administrative expense (at 10 percent)	$30	$30	$30
Net return, per unit	$8	$58	$68
Gross annual return	$29,000	$420,000	$820,000
Annual return on investment, percent	1	15	27

For example, someone on the board of directors might worriedly reflect, "Suppose only half the projected sales quantities is achieved and, further, that the actual manufacturing cost exceeds the estimates by 25 percent? After all, that's exactly what happened to us last year on Project XYZ." With such a pessimistic slant, a calculation shows that the first year would result in a loss, with the return rising only to 8 percent in the third year. At this rate of return, it would take from ten to twelve years just to recover the investment. Perhaps the board would finally decide that the truth could be expected to lie somewhere between the two extremes, with a pay-out time of six or seven years after introduction. If this view were to prevail, the project probably would be abandoned.

New products by acquisition

Many companies have grown very fast by acquiring other companies through exchanges of stock. Sometimes such mergers have proved highly advantageous, and sometimes they have been disastrous. Booze, Allen, and Hamilton state that less than two thirds of the acquisitions covered by their surveys have proved to be satisfactory to the acquiring companies.[8]

The motive most commonly behind an acquisition is the desire to obtain a new product line. Sometimes the "product" is only some patents and an experimental model, although the attractiveness of a proposed merger is greatly enhanced if a manufacturing and marketing capacity is also to be acquired.

The source of much unhappiness in corporate acquisitions is that realities do not always turn out to be so agreeable as the picture seemed to be during negotiations. Such a transaction is like that of the man who buys a box of apples, only to discover that all the good apples have been carefully placed on the top. But these situations can be avoided by using good business sense, and many corporations employ their new products departments to advantage in the careful and objective evaluation of potential acquisitions.

An interesting outline of some of the problems involved in creating new products has been provided by C. F. Rassweiler, vice-president of research and development for Johns-Manville Corporation. He says that a product in his field may take as long as seven or eight years from research discovery to commercial success, and that the decision on a new product will prove to have been justified:

If it can be achieved technically in the laboratory.
If it can be reproduced on a factory scale.
If after four years of research, the board of directors will approve the appropriation to build the plant.
If rejects and costs are reasonably low in factory production.

---

[8] *Management of New Products, op. cit.*, p. 15.

If consumers will buy the predicted volume of the product in six years.
If consumers will pay enough for the product to provide a satisfactory gross profit.

AND

If all this comes to pass before a competitor markets a similar product.[9]

---

[9] C. S. Rassweiler, "A Basis for Product Strategy Planning," in *Developing a Product Strategy, op. cit.*, p. 74.

NINE

Engineering management

In 1957, an English gentleman with the delightful name of C. Northcote Parkinson wrote a book called *Parkinson's Law.*[1] It has been rattling skeletons in administrative closets ever since. The book is more than funny: it it hilarious. But, as one reads, he experiences a chilling sense of recognition as it becomes apparent that Parkinson is very much in earnest.

Briefly stated, Parkinson's "law" is that *work expands to fill the time available for its completion.* This statement has serious implications for administrators. If there is insufficient work of a truly essential nature to occupy the attention of the work staff, then additional work will be generated to fill the void.

Parkinson concludes that administrations are bound to increase in size and that this increase is virtually independent of the actual work load. He cites some statistics to demonstrate his point, one example of which will suffice here. From 1914 to 1928, the British Navy declined by a third in officers and men and by two thirds in capital ships; during the same period, the Admiralty administration *increased* by nearly 80 percent.

Closer to home, an examination of American growth figures for the decade from 1947 to 1957 reveals that productivity increased at the rate of 3.3 percent per year, while the nonproduction work force increased at a much faster rate: 5.2 percent per year.[2]

---

[1] C. Northcote Parkinson, *Parkinson's Law* (Boston: Houghton Mifflin, 1957).

[2] D. W. Dobler, "Implications of Parkinson's Law for Business Management," *Personnel Journal*, January 1963, pp. 10–18. The productivity-growth rate given above is per *production* employee. An examination of productivity per *total* employee reveals a *decelerating* growth rate (from about 3.0 percent a year to 1.9 percent). However, it should be noted that the "nonproductive" work force includes many specialists (such as engineers), whose efforts are essential to industry. "Nevertheless," says Dobler, "the coincidence of these two situations—the expansion of the nonproduction work force and productivity's decelerating growth rate—should be sufficient cause for the individual practitioner to take stock of his own operations."

The question may be asked: "How does this pertain to engineering?" Just this: as soon as a company has more than a very few engineers on its staff, the need will be felt to create an organizational structure. Among the functions defined in an organizational plan will be many that are administrative in nature. Clearly, the manner in which the administrative structure is created and operated will have a considerable effect upon the overall efficiency of the engineering group.

Organizing for product development

Some engineers prefer to have no organization whatever (or at least they think they do, until the time when they become distressed by the lack of certain essential services). Others, if left to their own devices, will spend so much time in developing a smoothly functioning administrative machine that they have little time left over for actual productive work. Obviously, what is wanted is just the right amount of administration to provide the necessary support for the engineering activities, and no more. Many astute observers of the American industrial scene believe that we generally tend to err more in the direction of too much organization than too little.[3]

Two basic points are to be considered in building an organizational structure: 1) What is the job or mission to be performed by the group? 2) Which individuals *really* must have what kinds of information? In most organizations, a careful investigation can bring to light some kinds of activities that are not truly relevant to the basic mission of the group and can be dispensed with. Almost any group spends a certain proportion of its time preparing forms and reports—presumably for the information of others—and it is in this area that an investigation is likely to reveal activities that have no justifiable purpose. It has been observed that the most difficult thing to eliminate is a form that has acquired routine usage. As a typical case, one author tells of an instance wherein a form, reporting the number of square feet of drawings that had been completed by the group had to be filled out each month. Who needed this knowledge? A company statistician who had once been

---

[3] For example, see the following:

W. C. Lothrop, *Management Uses of Research and Development* (New York: Harper & Row, 1964). For 17 years, Mr. Lothrop was a scientific consultant and senior vice-president for Arthur D. Little, Inc.

P. F. Drucker, "Twelve Fables of Research Management," *Harvard Business Review*, January–February 1963, pp. 103–108. Mr. Drucker is a prominent management consultant.

P. Franken, "Research Inhibitions," *International Science and Technology*, May 1963, pp. 46–49.

caught short without that particular piece of information and was determined never to be caught again.[4]

If the two points of *function* and *who needs to know* are used as the guidelines in forming an organizational structure, then almost any scheme of organization can be made to work. It can readily be observed that identical forms of organization are, in some places, successes, and in other places, failures. Therefore, the specific form of organization itself must not be the "magic" ingredient. It is more likely that *people* will turn out to be the magic ingredient, if there is one.

If all the various forms of R & D organizations are sifted down to their essentials, three basic types will emerge: the *project* structure, the *functional* structure, and the *hybrid* structure.

The *project structure* (Fig. 9-1) corresponds in form to that of the decentralized corporation: each subgroup of the department is responsible for a complete project (or projects) and contains within itself all the functional competencies necessary to complete the projects. The major advantage of such a structure is that the boundaries of responsibility are crystal-clear; the major disadvantage is that functions are duplicated among groups.

The *functional structure* (Fig. 9-2) is highly centralized. A department is split into its functional specialties, and the functional subgroups operate on all projects passing through the department. One typical kind of functional grouping would separate the work into electrical design and mechanical design. Another kind would break the organization into aerodynamics, stress analysis, weights, materials, and test groups. Obviously, the kind of grouping will depend upon the branch of industry in which the company operates. The major advantage of the functional structure is that a greater technical competence can be achieved in the various engineering specialties than under the project structure. (It is more likely that the company will possess true expertise in material science, for example, if a group can focus all of its attention on this activity rather than be obliged to spread among many project groups.) The principal disadvantage of a functional structure is that it is difficult to pinpoint responsibility and certain important matters may get overlooked.

The *hybrid structure* combines the foregoing two forms and is characteristic of large companies. Functional groups exist, but each project is under the supervision of a project manager who shepherds his project through the various functional activities. Those who choose the hybrid structure generally do so in the hope that they will obtain the advantages of both the project form and the functional form. If they are successful in this aim, the hybrid form is the best of the three basic types. The

---

[4] V. Cronstedt, *Engineering Management and Administration* (New York: McGraw-Hill, 1961), p. 65.

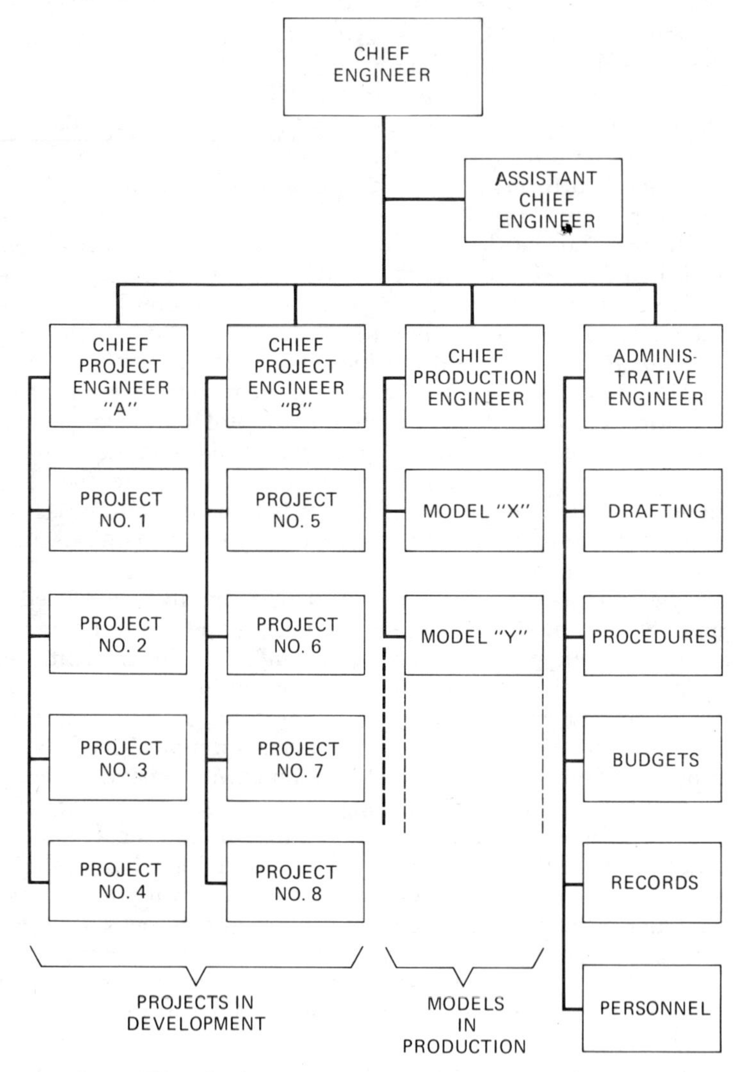

Fig. 9–1 Example of "project" organizational form.

danger is that, if the objective is not successfully attained, the result may be an achievement of the *disadvantages* of the other two forms—with none of the advantages.

There is yet another kind of functional structure, which is implicit in Figures 9-1 and 9-2. These illustrations imply that each project goes first through a "development engineering" phase and, then, through a "production engineering" phase and that these phases are carried out by different groups of people. The "development" phase is presumed to

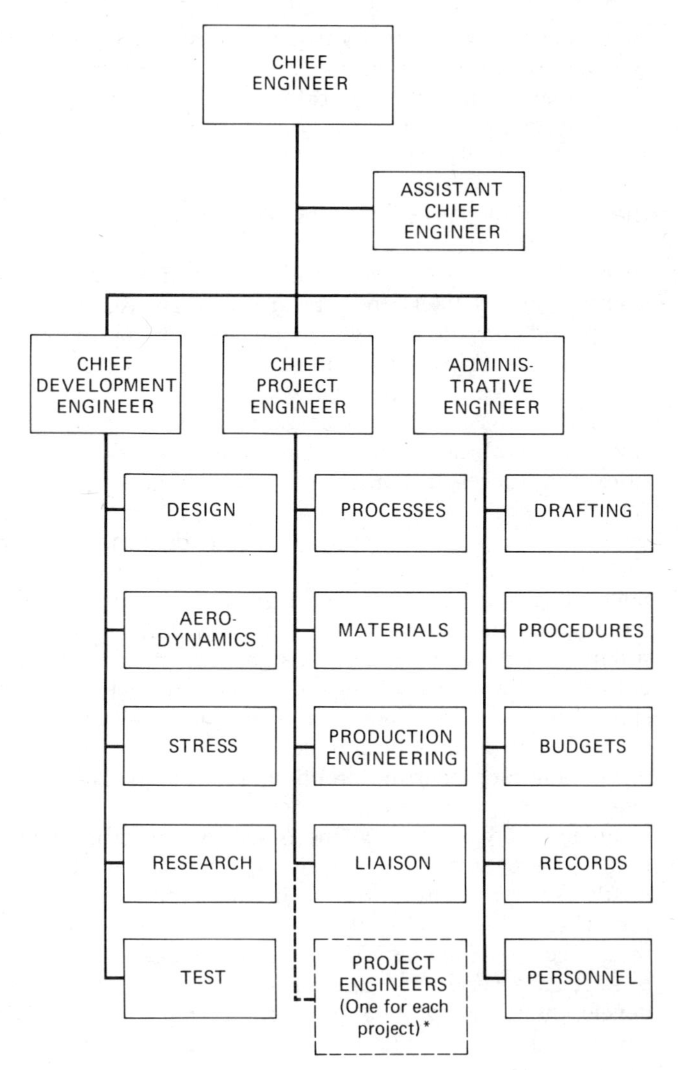

Fig. 9–2 Example of "functional" organizational form. * If there are no project engineers (block in dotted outline), the organizational form is known as "functional." If there is a project engineer for each project, the organizational form is known as "hybrid."

end with the demonstration of a successful prototype. During the "production engineering" phase, redesign for production economy is supposed to be accomplished, provided the function is not tampered with. Often, the "development" phase is performed at a research and development laboratory physically isolated from any of the producing divisions, whereas the "production engineering" is carried out by an engineering department that is an integral part of a producing division.

Many companies work identically in the fashion just described and work successfully, too. However, it is reasonably safe to say that nearly all these companies experience severe organizational problems during the process of passing projects from one group to the other. There is, for example, the peculiar factor known as "NIH." NIH means "Not Invented Here," which is another way of saying that one engineering group is highly likely to distrust the work transmitted to it by another group and may do the whole job all over again. Alfred P. Sloan, Jr., for many years president of General Motors, tells of problems of this nature involving conflict between the research organization and the producing divisions.[5] At least a part of the problem is semantic and involves the phrases "completion of development work" and "demonstration of a successful prototype." To the development group, the term "development of a successful prototype" may refer to a loosely assembled collection of bench-top apparatus that demonstrates an idea. Much of the novel design work can properly be considered completed at this point, but a great deal of design remains to be done; most of it is lacking in true novelty, but it is essential to the profitability of the product that the remaining design be done well. Much of the conflict between "development" groups and "production engineering" groups stems from a development group's tendency to think that a project in such a state is mostly completed except for the shouting, whereas a production engineering group will probably believe that it is hardly begun.

Some improvement can be made by charging each group to exhibit greater forbearance, but it is likely that much more progress will result from an alteration of organizational concepts. Figure 9-3(a) presents a diagram of the "sequential-project flow" approach, which might be regarded as the classical image of the manner in which projects are conducted and includes a transfer of the project from a development group to a production engineering group. Each step is begun after the previous one has been completed. The major disad-

---

[5] A. P. Sloan, Jr., *My Years with General Motors* (New York: Macfadden, 1965), pp. 71–94. Others have commented on the same subject. For instance, C. W. Perelle, president of American Bosch Arma Corp., says, "The first and foremost difficulty of adding to the [product] portfolio by this approach [acquisition of other companies] is that our engineers cannot accept anyone else's engineering." R. C. Clark, Jr., Manager of Research and Development for the Western Company, says, "Oddly enough, researchers resist new ideas (not their own) more than any other group."—from *Developing a Product Strategy*, E. Marting (Ed.) (New York: Amer. Management Assoc., 1959), pp. 79, 88.

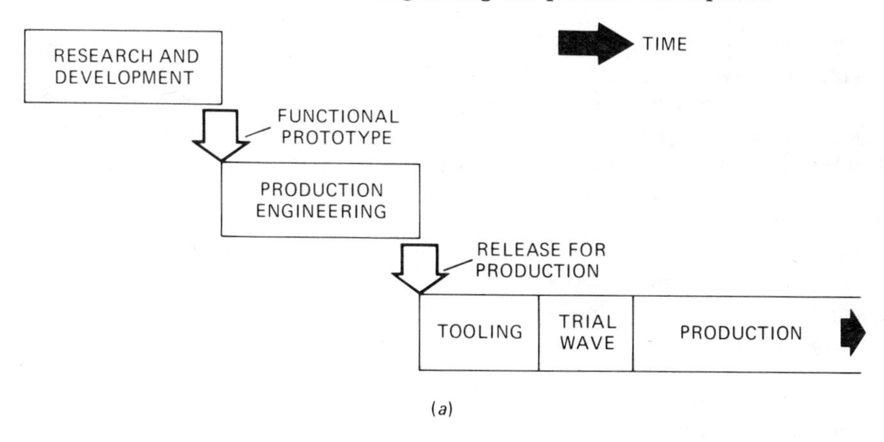

(*a*)

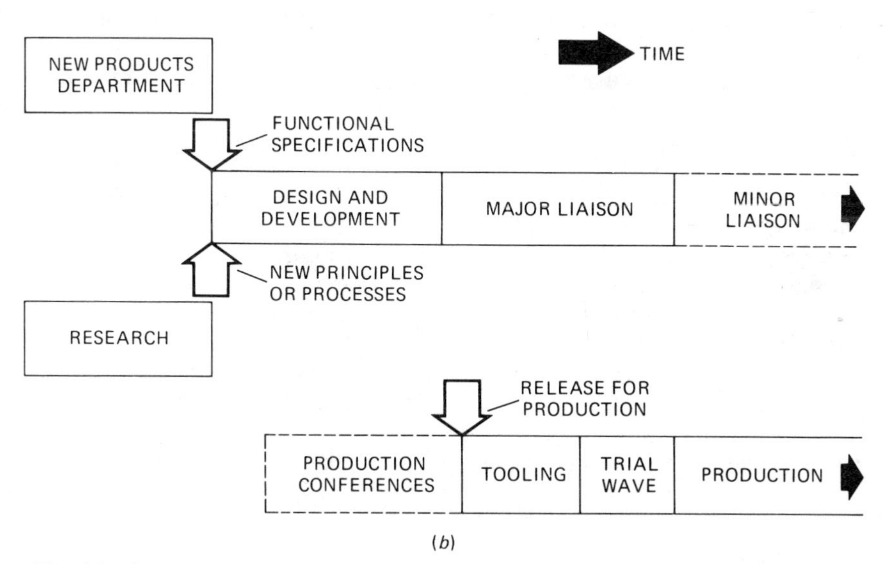

(*b*)

Fig. 9–3 Comparison of organizational approaches. (*a*) Sequential-project flow diagram. (*b*) Overlapping-project flow diagram.

vantage of such a procedure is that it consumes a great deal of time, partly because of lack of communication among groups. As a result, there is a great compulsion under modern-day competitive pressures to achieve telescoping of projects through overlapping of functions.

Such an overlapping is depicted in Fig. 9-3(*b*). Production planning begins even before the design is complete. In a limited way, some of the tooling may even be constructed before the design is finished so that parts made from actual production tools can be tested in the engineering prototypes. However, it should be apparent that such a course of events entails a considerable amount of risk.

One further aspect of an organization's structure, which seems to be considerably important, is whether it is relatively *closed* or relatively *open*. A *closed* system will typically exhibit a strong adherence to the usual organizational rules, with authority flowing from the top down along the lines of the organization chart and accountability for results flowing in the opposite direction. Great emphasis is placed upon productivity, upon enforcement of budgets and schedules, and upon going through proper channels. Even though such practices would seem only properly efficient and businesslike, it has been observed that some unexpected by-products can be high dependency (of individuals upon superiors), low autonomy, low opportunity for interaction, and low individual influence potential.

In the *open* system, greater emphasis is placed upon the autonomy of individuals and less reliance is given to achieving results by means of administrative control. It might be said that the open system emphasizes subject-matter, while the closed system emphasizes the methods by which the subject-matter is to be handled. In a closed system, new jobs tend to be adapted to the organizational structure while, in an open system, the structure is adapted to the jobs.

L. B. Barnes of Harvard made a study in depth of two engineering groups, one of which had a relatively closed system (Dept. "A") and the other, a relatively open system (Dept. "B"). Barnes cautioned that conclusive results cannot be obtained by examining only two groups, but did observe that Dept. "B" displayed higher individual satisfaction, less conflict, higher group performance, and greater individual opportunity than did Dept. "A."[6]

A recommendation in favor of an "open" system would seem to be equivalent to a vote for disorder, as opposed to order. Yet the implications of Barnes' study would seem to be that an excessively ordered environment may have an adverse effect upon productivity.

Manufacturing

If the work of the engineer is to reach the public, sooner or later it must be manufactured. It might be supposed that the responsibility of the engineering department essentially ends with the release of engineering drawings to the manufacturing department, but this is far from the case. In fact, during the early phases of manufacturing, it may seem to the participants that *nothing* goes right and that the engineer-

---

[6] L. B. Barnes, *Organizational Systems and Engineering Groups* (Cambridge, Mass.: Graduate School of Business Administration, Harvard University, 1960), pp. 149–152.

ing department is constantly embroiled in putting down crises. Everything that worked so well in the prototype model seems to go wrong in the production model.

There is a good reason for this, of course. First, mistakes are bound to occur. They must be found and corrected, and this is why it is important to initiate production with a "trial wave." (See Figs. 9-3(a) and 9-3(b).) Second, statistical variations in production parts unavoidably take place, and this will produce a certain amount of malfunctioning. Third—and probably most important—the transition from engineering to manufacturing is a transition from a closely supervised, one-at-a-time kind of activity to a widely dispersed mass production effort. The tasks must be broken down into thousands of subtasks, performed by as many different individuals, that are coordinated only by organizational procedures and pieces of paper; many difficulties will occur before this complicated system is running smoothly.

When a crisis arises on the production line, a frantic call is sent out for the engineers to come "put out the fire." Since everything is new and unfamiliar to the production people at this point, it becomes necessary to bring in the men who understand the product; and these are the men who designed it. In preparation for the start-up of manufacturing, a frequent practice is to assign a man to each project, in the very early phases, who knows his ultimate destiny is to become the "liaison engineer," that is, a man who really *knows* the product and can become the production line "crisis-stopper." During manufacturing start-up, this kind of arrangement helps to relieve the development people from demands that they drop their new projects and rush to put down crises. In any case, the early production period is likely to consume the attention of many engineers for a considerable time after start-up. Not until the occurrence of trouble has settled to a relatively low level, can the project be considered "complete."

Even after the subsidence of major crises, the engineering effort never entirely ceases on a product unless it is an extremely simple one. Tools wear and change, people are transferred, and new procedures are tried out. All of these may affect the product and probably will. Incomprehensible things like the following may occur, for example. A new tool is made to replace an old one, and in the process, it is discovered that the old tool was not producing parts that meet the dimensional specifications of the engineering drawing. Naturally, the new tool is made in correct accordance with the drawing, and then it is discovered that the new, "correct" parts don't function properly, whereas the old "incorrect" ones worked fine. Consternation ensues, and out goes the crisis-call to Engineering. Another "fire" must be put out.

On most mass production products, a perpetual cost-reduction program is under way throughout the product's effective life, and this requires the participation of the engineering department. The saving of a few cents on a part can add up to tremendous yearly savings. Elementary arithmetic will show that the saving of ten cents each on

200,000 parts a year can exceed a man's salary, and 200,000 parts a year is not very much in terms of mass production. However, in comparing two processes, the current one and a proposed one, it should be realized that *everything* is known about the current process—especially all the bad things; on the other hand, the things that are known about the proposed project will be predominantly good. It is practically axiomatic that unforeseen things will turn up as any new process is put into effect, and that these will mostly be developments that will increase the cost. Hence, any cost-reduction proposal must not only excel in thoroughness, but it also must offer an unusually good prospect for payoff, so that it can absorb all the unexpected setbacks and still prove worthwhile.

Dimensioning for the scrap pile

A subject for continual (and sometimes heated) debate, especially between engineering and manufacturing personnel, is the matter of *tolerances*.

The setting of tolerances is a very responsible matter and requires mature judgment: it is hardly something to be left to a drafting group, although some companies believe this is the correct thing to do.

Almost invariably, inexperienced engineers will set tolerances too tight. They do this partly because they lack knowledge of costs and partly because they sense that, in this direction, lies security. Some even exhibit an absurd tendency to specify all tolerances at $\pm$ 0.001 inches, simply because this is a convenient number.

Among manufacturing personnel, there is a tendency to believe that every tolerance is arrived at by Engineering through a comprehensive analytic process that will guarantee results if done correctly. This cannot be the case, of course. It is feasible and appropriate to analyze certain *critical* features of the design for their behavior under tolerance variations, but this is impossible for all the hundreds of thousands of dimensions involved in a complex product. Hence, most tolerances are established on the basis of judgment and experience. Again, the inexperienced (or the timid) will seek security in tight tolerances, but it is more advantageous to the company to find that magic condition wherein the tolerances are as loose as they possibly can be while they still permit the product to function properly. Tolerances that are tighter than necessary will only increase the number of parts which find their way to the scrap-pile. In fact, in some precision industries, it is taken for granted that extremely tight tolerances will be achieved only by virtue of a large amount of scrapped parts.

The use of supporting personnel

During past periods of acute engineer shortages, questions concerning the proper utilization of engineers have regularly been raised. Are engineers being used at their highest potential? Are many of them required to perform tasks that could be given, instead, to technicians or to clerks?

Statistics show that the ratio of technicians to engineers is far from that which is considered optimum. A survey conducted in 1963 disclosed a ratio of only 0.38 technician per engineer,[7] whereas most people seem to believe that 2.0 technicians per engineer would be desirable and some even favor a ratio of 4 to 1.[8] Among the various segments of industry, communications seems to be closest to the ideal, with 2.37 technicians per engineer.

What is a technician, and how may he be used efficiently? Broadly speaking, he is a technically trained person who works as a member of the engineering team in a support-capacity to the engineer, although he may possess certain skills which he has developed to a higher degree than has the engineer for whom he works. Such would be the case, for example, with the electronics technician who assembles experimental "bread-board" circuits, or with the mechanical draftsman who prepares crisp, workmanlike drawings from the engineer's sketches. The following list presents some kinds of activities in which engineering technicians engage:

Drafting
Estimating
Field Service
Inspection
Installation
Surveying
Technical Writing
Testing
Time Study
Tool Design
Maintenance[9]

Formerly, technicians just "grew" on the job, but there has been a strong movement, during recent years, to systematize the education of technicians. The Engineers' Council for Professional Development (ECPD), which is the accrediting agency for professional engineering education, also accredits programs for technician training. The majority of these programs are two years long (some are three) and result in the award of a degree such as Associate in Engineering, or Associate in Science.

From the viewpoint of engineering management, all jobs that can be efficiently assigned to a technician should be so assigned. This seems to be a simple enough imperative; yet it is often violated. Strangely, the utilization of technicians is sometimes resisted by the

---

[7] *Demand for Engineers, Physical Scientists and Technicians*—1964. (New York: Engineering Manpower Commission, EJC, 1964), p. 51.

[8] *Engineering Manpower, A Statement of Position.* (New York: Engineering Manpower Commission, EJC, 1963), p. 23.

[9] *Technician Career Opportunities in Engineering Technology.* (New York: ECPD).

individual engineer. Some engineers gain personal satisfaction from working with their hands: wiring their own circuits, making their own drawings, or running machine tools in the experimental shop. From management's point of view, such practices are hardly justifiable because: 1) an engineer will seldom be as skilled as the technician who is a specialist in any of these fields, and 2) the engineer is being paid a salary higher than the technician's to do work that the technician could do better.

There are other—and more powerful—reasons why the technician-engineer ratio may be less than optimum. The main cause is simply that technicians are in even shorter supply than engineers. Furthermore, it looks as if this situation will continue unless there is a dramatic increase in technician enrollments. In recent years, this country has been producing less than one technician for every two engineers; this ratio is barely enough to maintain the status quo.

Another reason for less-than-optimum technician–engineer ratios in some companies can be traced to layoff practices during periods of cost-cutting. When an engineer is laid off, a lot of know-how goes with him, and this know-how can be reacquired later on, only with much difficulty. Because there is a much smaller know-how investment in a technician, there is a natural tendency to let a high proportion of technicians go during a layoff and to keep the engineers. Naturally, the result of such a practice is that the remaining engineers must pick up the load formerly carried by technicians.

Many fine points of judgment are involved in deciding what should be handled by engineers and what should be handled by technicians. Often, technicians are capable of discharging some of the simpler phases of design and should be encouraged to do so in the interests of efficiency. However, various responsible groups have expressed alarm over the tendency of many engineers to avoid design of a detail nature and, consequently, to allow responsibility for such work to be passed by default to technicians. In 1959, a committee appointed at the Massachusetts Institute of Technology to study engineering design commented that many young engineers seem to believe any problems not requiring higher mathematics are beneath their dignity and so let them pass to technicians. "This attitude," said the committee, "often prevails in spite of the clear indication that the most important decisions in a design problem must often be made without assistance from higher mathematics."[10] The main inference to be drawn from the foregoing is that mature judgment is necessary to decide what should be left to technicians and what should not, and that mathematical content is not necessarily a good criterion.

The concern about detail work is not confined to the United States.

---

[10] "Report on Engineering Design," *Journal of Engineering Education*, April 1961, pp. 645–656.

The United Kingdom Atomic Energy Authority in 1963 commented, "There is a tendency for designs to be spoiled by lack of attention to detail and this often causes difficult and costly rectification work at site that could have been avoided by more rigorous thought at the design stage. . . ." According to a British committee appointed to study such matters, one reason for this state of affairs is the unwillingness of qualified design engineers to work on detail design.[11]

If there is a moral to be drawn from the preceding section, it is that engineers can profitably use technicians as assistants but cannot abandon responsibility for a design matter to them just because it happens to be of a detail nature. As the British committee on design said, "In design *everything* matters."

Metrication

The issue of conversion to the metric system has faced the United States increasingly, as the twentieth century has progressed. America "officially" adopted the metric system in 1889, but it never came into wide use in this country except for a few items such as ball bearings, lenses, and film. In the early 1900s, some highly emotional battles were fought over conversion, between pro-metric and anti-metric forces, but two wars and a lengthy depression caused the country to lose interest in the matter until the 1960s, when its dominance of world markets began to wane and metric conversion again became an issue. As of about 1971, it appeared that the United States at last was going to have to yield and go metric, largely because it was the only industrialized nation in the world that had not already done so.[12]

Most of the supposed "natural" advantages of the metric system are illusions. For example, we are told that it is a simple system requiring conversion factors which are all powers of 10. But this claim is not true. It overlooks the fact that the hour is divided into 3600 seconds, whether in the metric or the "English" system. Hence, one watt-hour equals 3.6×10^3 joules. (Reminder: a watt is defined as one joule per second). Also, one calorie (amount of heat required to raise one gram of water 1°C) is equal to 4.184 joules. Most other metric conversion factors do manage to work out as powers of 10, although additional troubles can develop if the acceleration of gravity gets involved, because it inconveniently comes out to be 9.81 m/sec^2, at least at the earth's surface.

---

[11] *Engineering Design, Report of a Committee Appointed by the Council for Scientific and Industrial Research to Consider the Present Standing of Mechanical Engineering Design* (London: Dept. of Scientific and Industrial Research, Her Majesty's Stationery Office, 1963), p. 16.

[12] D. White, "Metrication: An ASME Report to the Membership," *Mechanical Engineering*, March 1973, pp. 69–75.

Many promoters of metric conversion have castigated the United States for using a measuring system consisting of inches, feet, yards, rods, and miles, with the accompanying peculiar conversion factors of 3, 12, 16.5, 5280, and so on. Probably such an irrational system should properly be ridiculed, but the critics overlook the fact that the system of feet and inches, while widely used in the home, is hardly used at all in industry, except in construction, yardage goods, and a few other areas. Most of American manufacturing industry has for years used the decimalized inch as its standard. All measurements are in inches—even long measurements, like the wingspan of an airplane. No fractions are used; decimal equivalents are used instead. Many of the supposed advantages of the metric system are in actuality advantages of decimalization, and manufacturing industry in the United States has been decimalized for a long time.

It has been supposed that in converting to the metric system we would adopt the cgs (centimeter-gram-second) system, or perhaps the mks (meter-kilogram-second) system, the latter of which is now called the "Système International," or simply SI. Both of these systems are called "absolute" systems, because the unit of mass is a defined quantity, and force is a derived quantity. The cgs system defines the gram as a unit of mass, and the dyne as a unit of force, while the mks system uses the kilogram and newton, instead. Most engineering students in the past managed to ignore the cgs and mks systems as soon as they completed their physics and chemistry courses, opting instead for the pound as the unit of force and carrying along a funny unit of mass called a "slug," whenever it seemed necessary to do so. If any of these engineers remembered there was an "English" absolute system in which the pound was a unit of mass, and something called a "poundal" was a unit of force, most of them managed to keep from using it.

In converting to the metric system, we were plunged directly back into this confusion of units. Several technical magazines in the United States converted to the metric system in the early 1970s, and readers had to begin to puzzle out quantities in unfamiliar units. For example, pressures might appear in terms of N/m^2, and even many well-informed persons had to struggle a bit to convert the quantities into lb/in^2, discovering meanwhile that the acceleration of gravity kept intruding itself into the conversion process. Then, the astonishing revelation came from knowledgeable international engineers that engineers in metric countries of Western Europe did not themselves actually use the SI! They used metric units all right, but instead of using the newton as the unit of force, they used the kilogram force unit, in an exact parallel to the long-established United States/British custom of defining the pound as the unit of force, and jettisoning the poundal. One international engineer claimed, "Although most enlightened European engineers know of the existence of the Newton force unit, hardly anyone uses it . . . the claim that to support one liter of water against gravity requires a force of 9.81 newtons, rather than simply one kilogram of force, does elicit some chuckles."[13] It appeared that one unexpected consequence of metrication was that engineers would have to go back to their college physics texts and bone up on "absolute" and "gravitational" systems of units.

The basic issue in metrication really does not involve any supposed advantages of the system itself, but the fact that the world is on two different measuring standards. If it were only a matter of making conversions from millimeters to inches and vice versa, the whole matter would not be of much consequence. But the problem is that threaded parts from the two systems do not fit together, tools made under one system do not fit on parts made under the other, and standards in general under the two systems are not compatible. It is all a matter of dollars and cents. To make the conversion will cost the United States a lot of money—some have estimated it will cost $50 billion.[14] Obviously, this is an unpleasant economic pill for the country to swallow and is precisely the reason the United States has resisted conversion so long. But others believe the failure to convert will cost the country much international business, because of the problems with interchangeable parts, and it is this factor which is driving America to convert.

---

[13] "50 Kg of Potatoes Supported by 490 N," *Mechanical Engineering*, October 1973, p. 78.

[14] "Metric Conversion: A Triumph of Technology over Common Sense?" *Professional Engineer*, September 1973, p 51.

TEN

Salaries and other rewards

The rising curves

It has been observed that salary almost never appears at the top of the list when engineers are asked to indicate those things that are important to them in the ideal job. Instead, it modestly shows up in third or fourth place. For example, in one survey, the leading desirable job characteristics were reported to be 1) opportunity for advancement, 2) creative, challenging work, 3) good salary, 4) recognition of achievement, and 5) the chance to keep up with new developments.[1]

Some have been tempted to believe that this means engineers have little interest in money. However, it is more likely that most engineers simply have not experienced any need to worry about salaries because they have been pretty much in a "seller's market" ever since 1950, with the exception of the period of the 1969–1971 recession.

Although salary is well down the list, it will turn up in the number one spot any time a man believes that his salary is not being handled in a fair manner. This usually means that he expects his employer to keep salaries in step with "what is going on outside."

The widespread distribution of salary surveys makes the income of engineers among the best documented of subjects. Surveys are regularly taken by all kinds of groups, including manufacturers' associations, regional organizations, engineering societies, and the U.S. government. Among the better-known surveys are those of the American Management Association (Executive Compensation Service), the National Society of Professional Engineers, and the Engineers Joint Council (Engineering Manpower Commission, EMC).

At one time, a criticism of salary surveys was that they lumped all engineers together, regardless of their responsibility levels. Former engineers who had become company executives were still included in

---

[1] *Career Satisfaction of Professional Engineers in Industry* (Washington, D.C.: The Professional Engineers Conference Board for Industry), p. 20. In cooperation with the National Society of Professional Engineers.

the surveys and acted as average-raisers. But, since 1964, the Engineering Manpower Commission has divided its report so that the incomes of supervisors and nonsupervisors are separated. EMC's survey is concerned with degree-holding engineers only, and usually includes on the order of 200,000 engineers.

Figure 10-1 shows that the average dollar increase per year for a "median man" averaged about $820 per year during the period 1953 to 1972. Following 1965, the curves display a steeper trend than in prior years, but the trend is largely offset by inflation. Figure 10-2 shows the inflationary effect. Engineers were making steady gains in real income from 1953 up until about 1968, but after that time inflation overtook them, erasing any salary gains, and even causing a slight decrease in real income. Mostly, this effect would seem to be the result of combined inflation and economic recession, but some members of the profession chose to interpret the lack of gain in engineers' salaries as a sign that engineer shortages were a thing of the past.

Figure 10-3 displays the customary "maturity curves" obtained by plotting engineers' salaries against the years they have held their bachelor's degrees. The labels "10th, 25th, Median, 75th," and "90th" refer to percentile levels and mean that 10, 25, 50, 75, and 90 percent, respectively, of the persons in the survey are below the curves indicated.

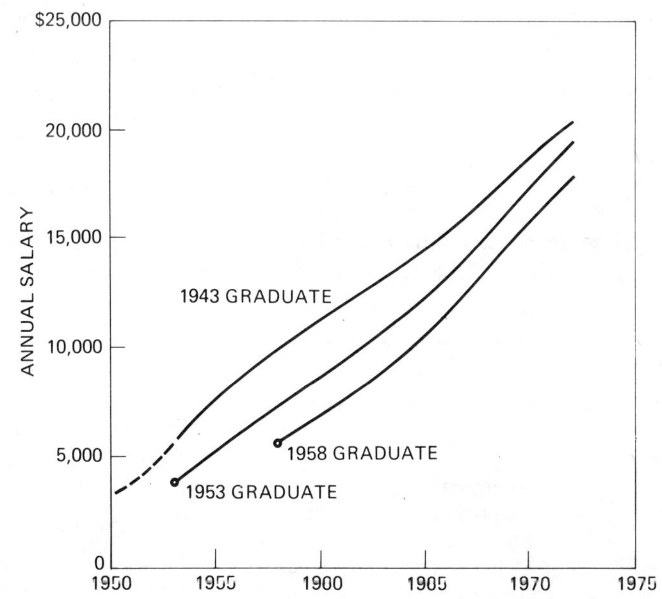

Fig. 10–1 **Typical rates of salary increase for three hypothetical engineering graduates, based upon median salaries reported. (Source: *Professional Income of Engineers, 1972.* New York: Engineering Manpower Commission, December 1972.)**

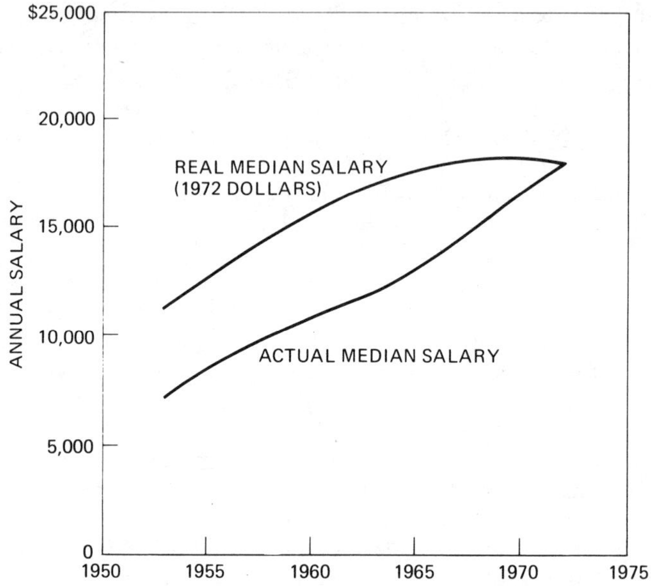

Fig. 10–2 Change in median salaries for engineers, from 1953 to 1972, for engineers with 15 years of experience. "Real" salary curve represents adjustment for 36.2 percent increase in consumer price index. (Source: *Professional Income of Engineers, 1972*. New York: Engineering Manpower Commission, December 1972.)

The median salaries in 1972 in different industries for engineers with 15 years' experience were:

	Nonsupervisors	*Supervisors*
All engineers	$17,500	$20,000
Aerospace	18,000	21,000
Chemicals	18,000	20,500
Electronic equipment	20,000	22,000
Machinery	16,400	19,500
Construction and consulting	16,400	19,500
Research and development	19,100	24,000
Federal government	18,400	21,000
State government	16,400	16,000

Source: *Professional Income of Engineers, 1972* (New York: Engineering Manpower Commission, December 1972).

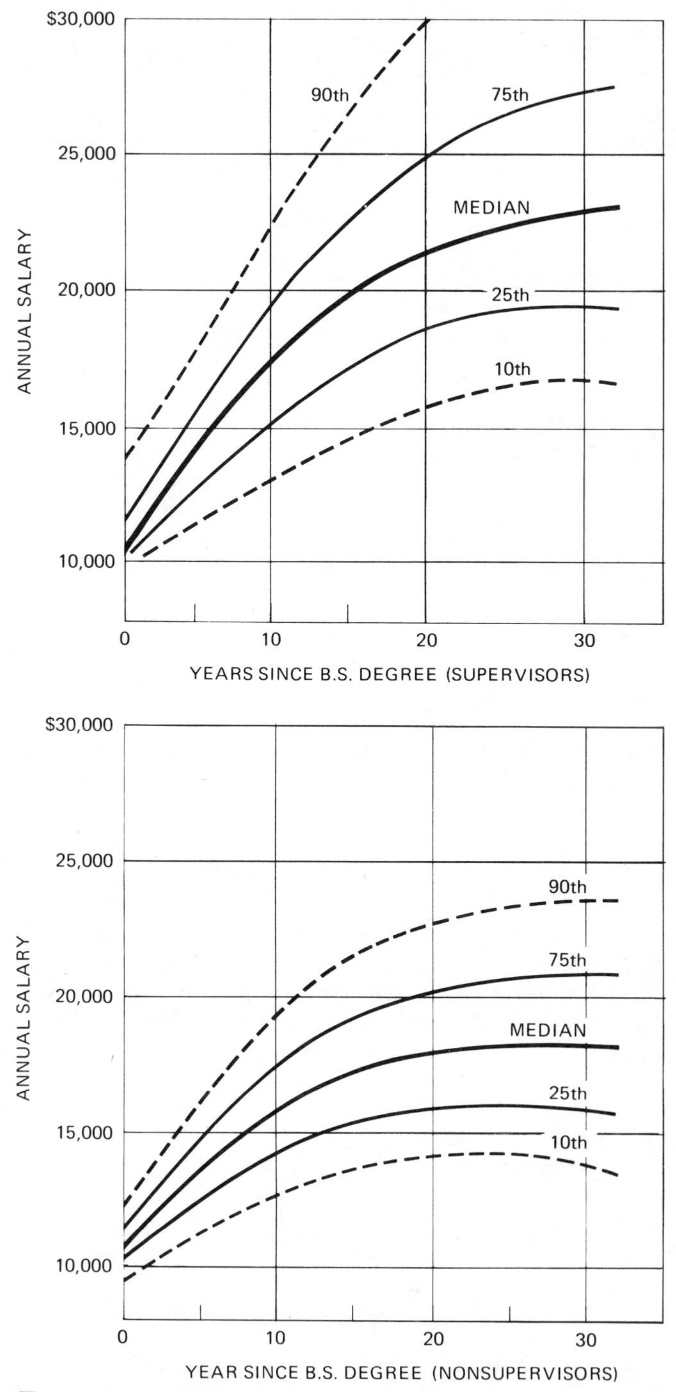

Fig. 10–3 Salary curves for U.S. engineers, 1972. (Source: *Professional Income of Engineers, 1972*. New York: Engineering Manpower Commission, December 1972.)

It can be seen that the typical increment for supervisors is on the order of $2500 to $3000. The apparent reversal of supervisory and nonsupervisory salaries for engineers in state government is probably a statistical accident, and would not be expected to prevail in a given state.

Starting salaries

For information concerning employment trends for college graduates, Dr. Frank S. Endicott, director of placement for Northwestern University, Evanston, Illinois, has no peer.[2] Since 1947, Dr. Endicott has conducted regular surveys on employment trends. Figure 10-4 shows the movement in starting engineering salaries since 1947, which has been steadily up, except for the 1949–1950 and 1970–1971 periods. From 1947 to 1972, starting salaries for engineers increased by a factor of 3.7, during a time when the cost of living approximately doubled.

The observation has been made that starting salaries have moved upward faster than salaries for more experienced men. The name "compression" has been given to this phenomenon. However, Figures 10-2 and 10-4 show the following: in the 10-year period from 1962 to 1972, starting salaries increased a total of about $4000, but the median salary level for 15 years' experience rose by $6300. Thus, although, on a percentage basis, starting salaries rose a greater amount (since they started from a lower base), there was no "compression" on a dollar basis.

Do bricklayers really earn more than engineers?

Many engineers believe that construction workers are paid more for their efforts than engineers and that the condition constitutes a gross injustice. They argue, with some logic, that a long and arduous college education should produce better financial results than this. But both the logic and their sense of injustice are wasted, because the premise is false.

It is easy to be misled. For example, the "Construction Scoreboard" in *Engineering News-Record* tells us that bricklayers in 1972 were making $9.05 per hour.[3] The engineer quickly multiplies (taking into consideration that an average month contains 4.33 weeks, not 4) and concludes that the average bricklayer in 1972 was making $1570 per month, or almost $19,000 per year. Since the median income for

---

[2] See F. S. Endicott, *Trends in Employment of College and University Graduates in Business and Industry, 1973* (Evanston, Ill.: Northwestern University).

[3] *Engineering News-Record*, July 6, 1972, p. 41.

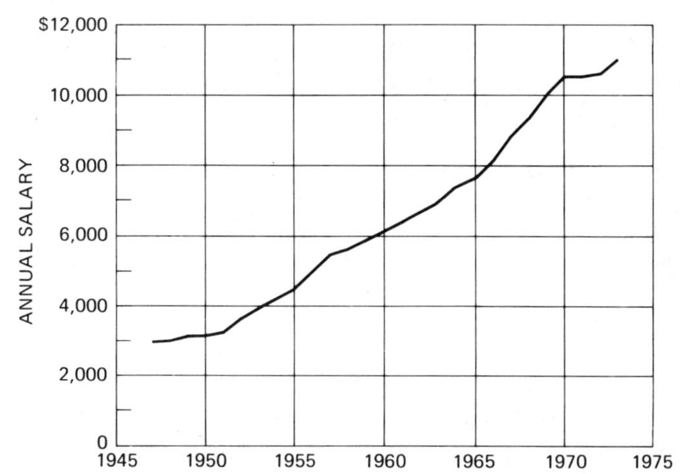

Fig. 10–4 Beginning salaries, from 1947 to 1973, for graduate engineers without experience. (Courtesy of F. S. Endicott Surveys, Northwestern University, Evanston, Ill.)

graduate engineers (nonsupervisory) in 1972 was about $18,000 per year, he decides that something is seriously wrong.

The vital missing ingredient is steadiness of employment. Most construction workers miss much working time because of such things as adverse weather and variability of construction contracts. The median annual earnings for various occupations reported by the U.S. Bureau of the Census in its report for 1971 were:

	1971 Annual Earnings (Median)
Craftsmen, in construction	$ 8,291
Foremen	10,826
Physicians and surgeons (self-employed)	25,000
Physicians and surgeons (salaried)	18,096
Engineers (graduate and nongraduate)	14,029
* * * * *	
Engineers (graduate, 15 years' experience, from EMC survey; see Figure 10-2)	17,500
Engineers (graduate, no experience, from Endicott survey; see Figure 10-4)	10,500

Source: *Money Income in 1971 of Families and Persons in the United States.* (Washington, D.C.: U.S. Bureau of the Census, December 1972).

An interesting observation concerning the foregoing table relates to income for physicians and surgeons. A widespread belief exists that members of the medical profession regularly earn salaries of $50,000-plus, perhaps up to $100,000 and more. No doubt some physicians do earn salaries of this size, and these cases are the ones that attract attention. The median earnings of self-employed physicians is considerably greater than that of engineers, but the median earnings for salaried physicians is not much greater than the median for graduate engineers. This condition exists in spite of the fact that physicians spend many more years in university training than do most engineers.

Salary administration

Salary administration is a difficult, and sometimes frustrating, activity. Many employers expend considerable effort and expense in the attempt to analyze salary structures scientifically and to carry out carefully planned review systems. All this is done in a genuine attempt to establish a remuneration for each person that is commensurate with his contribution.

One of the more difficult matters handled by every salary program is determining the basis upon which a man is, or is not, to receive a raise. It is generally considered undesirable for a man to receive raises automatically as a function of length of service, except perhaps in the first year or so following employment. Instead, increases usually are granted on merit.

However, it is often difficult to decide exactly what is meant by "merit." One interesting salary program that handles this question in a straightforward manner is Union Carbide's.[4] This program calls for supervisors to make a forced choice, concerning each employee's standing among the categories *satisfactory*, *commendable*, and *outstanding*. (Presumably, if a man is classified *unsatisfactory*, he is terminated.) In 1964, a "satisfactory" man was given an increase of 3 percent, a "commendable" man received a 5- to 8-percent increase, and an "outstanding" man was raised by 12 percent or more. It was stated, "The satisfactory increase percentage is usually designed to correspond with the annual formula increase." (In 1964, the formula increase called for all salary ranges to rise by 3 percent.)

The truly outstanding man is no problem to handle under a salary administration program. Neither is the "commendable" man. Salary increases for both usually can be easily justified on a clear merit basis.

---

[4] C. S. Dadakis, *Job Evaluation and Salary Administration for Engineers* (New York: American Society of Mechanical Eigineers, ASME Paper No. 64-MD-24, 1964).

However, a glance at Figure 10-3 reveals a distinct flattening in the salary curves of the "median man" after his twentieth year, when he is only halfway through his 40 years or so of productive life. This means that, if it were not for the existence of upward-pushing forces such as inflation and continuously rising engineering salaries, our fictitious "median man" would reach a salary plateau halfway through his career and would never again get another increase.

There appears to be a strong interest in a type of plan known as the "curve approach." It is a controversial one. One proponent frankly admits, "Many compensation managers damn it; many others have high praise for it."[5]

In such plans, three elements are allowed to influence salary growth: educational attainment, maturity factors, and job performance. The rationale for including maturity factors is that an increase in years of experience implies an increase in effectiveness and judgment and, further, that a man is entitled to salary growth as a reward for satisfactory performance. The third factor is included so that truly superior performance can be recognized by larger salary increases. One required step in the administration of these plans is that every employee be ranked on the basis of performance on a list, sometimes referred to as a "reverse layoff list." This list is then used to justify increases given for superior performance.

Proponents of the curve approach maintain it is more successful than other plans in achieving the basic objectives of all such programs, which are to reward outstanding performance and to establish equity of compensation among all employees, as far as this is possible. They also maintain that compensation plans based on the curve approach are simpler to administer.

In their efforts to build rational salary administration programs, companies that do not adopt the curve approach usually start from zero and attempt to build up a point value for each job, based upon its worth to the organization. Typical criteria would be the amount of knowledge or skill required, the degree of original thought and action necessary, and the extent of accountability and responsibility. These are used in a comparative fashion against other jobs in the organization to establish a point-rating. They are also used to prepare job descriptions.[6]

Once a set of relative point values is obtained, it might be supposed that salary ranges, based upon these points, could immediately be established. However, the influence of the marketplace upon salaries

---

[5] E. A. Shaw, "The Curve Approach to the Compensation of Scientists," in *The Management of Scientific Talent*, J. W. Blood (Ed.) (New York: American Management Assoc., 1963), p. 147.

[6] C. W. G. Van Horn, "Compensating Scientific Personnel," in *The Management of Scientific Talent, op. cit.*, pp. 130–132.

for different kinds of jobs must be taken into account. This means that salary surveys must be consulted to ensure that rates will be competitive.

In salary administration, there may be a temptation to establish rate ranges *a little higher than the market*, to give a recruiting edge. Obviously, if all companies do this, the result is a bootstrapping operation; this effect alone would be enough to keep salary curves rising. Even if only a few companies do this and all the others merely aim at the median, a net upward force is still generated.

It is an unusual company that publishes its rate ranges for all to see. (A survey by Booz, Allen, and Hamilton produced the remarkable information that 80 percent of the companies in the survey did not have written salary policies available to their technical personnel.[7]) Generally, a man is told his own range and, probably, he will be told what is necessary before he can advance into the next range. Information concerning salaries paid to specific individuals is almost universally considered confidential.

A basic ingredient of most salary programs is a provision for regular reviews of performance. If the results of surveys are to be believed, the manner in which many programs are carried out must be poor.[8] Some companies do a consistent job of handling performance reviews, but in many others, these activities are conducted sporadically, if at all. In some cases, if the man actually is to receive a raise, he is given a review by his supervisor, but if no raise is forthcoming, the matter of a review is ignored. Many times, the excuse is given that time is insufficient for periodic reviews, or supervisors may express doubt that higher management is really interested in having them take the time.

Situations such as the following are nearly independent of any salary programs that may be in operation:

1. The rate at which an engineer just out of college advances in salary is almost independent of his performance. If he is judged sufficiently valuable to continue on the payroll at all, his increases in the first year or two will approach the point of being automatic. After his second year with the company, his individual performance will begin to show its influence, and the truly outstanding man can move ahead very fast. Even during the first five or ten years, a competent man can expect a more or less steady salary growth. After the tenth year, the salaries of many will begin to level out.

2. An individual's future salary history will be strongly influenced

---

[7] J. L. Wyatt, "Are Creative People 'Different'?" *The Management Review*, July 1959, p. 21.

[8] J. W. Riegel, *Administration of Salaries and Intangible Rewards for Engineers and Scientists (Part 1)* (Ann Arbor: University of Michigan, 1958), pp. 63–66.

by the salary at which he was hired.[9] As a general rule, no matter what kinds of positions eventually become open to a man, each increase will be considered in the light of what he is currently making.

3. A curious feature of salary curves in some surveys is that they turn downward at about the twenty-fifth year. (Note the 10th and 25th percentile curves in Figure 10-3, for nonsupervisory engineers.) This does not necessarily mean that men have been forced to take salary cuts as they grow older (although such a thing could happen to a man if he changed jobs late in his career). Instead, the "drooping" appearance of the salary curves probably means that the middle regions of the curves have been pushed up faster than the right-hand portions. This signifies that companies sometimes consider their younger men more valuable than their older engineers, which opens another question: "Why should companies feel this way?" A study of 2500 engineers by Dalton and Thompson provides some of the answers.[10] Technical obsolescence is part of it: many of the older engineers have simply failed to keep up with their profession. Another factor is that many of the more capable engineers have been moved into management, although this is held to a relatively minor factor. Probably a more important reason is that companies are worried about the possibility that the younger engineers may leave them, and so tend to increase their salaries more readily than for the older, less mobile engineers. Another negative factor is held to be the kind of rating system that automatically signals to half the people on the work-force that they are below average. Such a system may involve negative reinforcement and cause those rated below average to become discouraged.

Rewards other than salary

It can hardly be questioned that there are other than monetary rewards in a job. Here are some comments from practicing professional engineers, about engineering as a career:

> The major thing is the creative aspect. There is a personal satisfaction in seeing a job completed.
>
> Mainly, there is the feeling of accomplishment, the challenge. . . .
>
> . . . there is considerable variety in the work.[11]

---

[9] R. E. Walton, *The Impact of the Professional Engineering Union* (Cambridge, Mass.: Div. of Research, Graduate School of Business Administration, Harvard University, 1961), p. 64.

[10] G. W. Dalton and P. H. Thompson, "Accelerating Obsolescence of Older Engineers," *Harvard Business Review*, September–October 1971, pp. 57–67.

[11] *Career Satisfactions of Professional Engineers in Industry, op. cit*, p. 11.

J. W. Riegel found the following intangible rewards to be the most important to the engineers interviewed in his survey (salary was deliberately excluded):

1. Challenge and variety in the work
2. Having ideas accepted and put to use
3. Treatment as a professional: status and personal freedom
4. Recognition of contributions by higher management
5. Association with able professionals
6. Opportunities to learn[12]

Job security did not seem to rate very high on the lists—a finding which might have come out differently if the survey had been conducted just after the 1969–1971 aerospace cutbacks.

Other intangible rewards mentioned by survey participants include being treated as part of management, having good clerical assistance available, and enjoying reasonably private working quarters. The preceding do not place very high in engineers' specifications of the ideal job. Yet, when the survey participants were asked what factors caused dissatisfaction because they were *lacking* in their own jobs, the factors just listed were at the top of the list.

An interesting type of reward is the bestowing of titles in lieu of raises. Even though this classic situation serves as the butt of countless jokes, it is unlikely that it often takes place for the implied reason. The giving of important-sounding titles is a form of intangible reward essentially independent of, and in addition to, the giving of raises. The justification usually offered is that an impressive title helps a man in dealing with outsiders. A more realistic reason is that it is another way of satisfying the need of an individual to show growth in his career. The only real danger in the practice is when it extends so far as to confer managerial-sounding titles upon positions that have no managerial content whatsoever.

Many a writer has lamented that an engineer-turned-manager is an engineer lost to engineering. Yet, it has also been pointed out that promotion into management has long been the only method of further advancement open to technical people. Some companies have attempted to alleviate this situation by offering dual promotional ladders: one ladder leads into the conventional management route; the other provides for salary levels equal to those of middle management, but for outstanding scientists or engineers who do not have administrative responsibilities.[13] That this approach has limitations is put into words

---

[12] Riegel, *op. cit.* (*Part* 2), p. 6.

[13] D. S. Beach, *Personnel: The Management of People at Work* (New York: Macmillan, 1965), p. 697.

by an engineer in one survey, who said, ". . . there are more steps available in the managerial direction, and they lead all the way up to the president of the company. Obviously, the technical steps available stop far short of that."[14] Nevertheless, programs of this type have done much to relieve former distress and anxiety. Many promotable engineers, who really don't want to be managers, are thereby given considerable extra salary potential.

---

[14] *Career Satisfactions of Professional Engineers in Industry, op. cit.*, p. 16.

ELEVEN

Creativity

The trouble with words

The word "creativity" is in a rather curious state: nearly everybody recognizes creativity when he sees it; yet nobody can define the word in a fully acceptable manner. From such a simple cause stems much bitter controversy. According to the dictionaries, to "create" means to bring into existence something that did not previously exist, but this definition satisfies almost no one. Therefore, this entire chapter may be regarded as an attempt at a definition.

In any attempt to discuss "creativity," another term, just as troublesome, almost immediately presents itself: the term "invention." Every layman is instantly confident of the word "invention." He knows exactly what it means; he can think of countless "inventions": the electric light, the safety pin, the phonograph, even the atomic bomb. It is only Patent Office examiners and United States Supreme Court justices who believe there is a problem in defining what an invention is. The legal problems surrounding this word are severe and will be discussed in Chapter 13.

For the purposes of this book, an invention will be regarded as something that is clearly possessed of novelty and usefulness, has identity as a distinct device, process, or system, and has been "reduced to practice." (Even though an idea may have been reduced to practice, it still may not be commercially practical. "Reduction to practice" may consist of nothing more than a written description in a patent, or it may consist of a series of laboratory demonstrations that function only under carefully controlled conditions. The long, and usually painful, task of refining an idea until it reaches the point of commercial realizability is known as "development.")

It is almost impossible to talk about creativity without also talking about patents. The aforementioned layman, who had no trouble with the term "invention," would probably also assume that "creativity," "invention," and "patent" all refer to essentially the same thing. A slightly better-informed layman would know this is wrong, for he could immediately point out that Einstein could not have patented his theory of relativity, even though the highest possible order of creativity was involved in bringing it into being. Nor would Einstein's theories normally be termed "inventions." As another example, the Golden Gate Bridge represents creative civil engineering of the first magnitude, but it is not an invention, nor is it patentable, because suspension bridges have long been known.

Most important of all to the present discussion: the innumerable innovations and design decisions which go into the final embodiment of a modern-day device (whether it be an automobile, a chemical plant, a satellite, or a zipper fastener) represent acts of creativity of varying degree. Some of these innovations may be patentable and some may not. However, they are all acts of creativity, and it is with creativity *in this larger sense* that we are concerned.

Some people will be quick to point out that there are other kinds of creativity than have been mentioned in the preceding sections. For example, authors, artists, and composers are just as concerned with creativity as are engineers and scientists. It may even be that the mysterious fundamental sources of creative potential are the same, whether they are expressed in technological or nontechnological forms. Humanity has an extremely high regard for artistic creativity, but it will not figure in this discussion, except as it impinges upon engineering in the form of the special activity known as "industrial design."

Differing kinds of technical creativity

In the previous section Einstein's theories were given as an illustration of the highest type of creativity. There might be a temptation to rank all new scientific knowledge in this highest category; however, some kinds of scientific effort do not involve much creativity, since they consist of applying a stimulus and observing an outcome. In true scientific creativity, there is a recognition of novel potentialities and a new ordering of the knowledge that permits significant and enlightening generalizations (in other words, the statement of a theory).

For instance, the formulation of Newton's laws and that of the laws governing the behavior of charged particles fall into this highest category of creativity. In each of these accomplishments, the establishment of a set of mathematical relations between observed experimental results constituted the central act of creation.

The cyclotron is a very interesting example of creative activity because it necessitated that a scientist, E. O. Lawrence, do some engineering (that is, design the cyclotron) in order to continue with his scientific research concerning high-energy particles. It is unlikely that anyone but a nuclear physicist would ever have conceived of the cyclotron, both because it would require a person of such background to recognize the need and because it would require training in nuclear physics before one could know how to go about making it.

Magnetic recording is a kind of example which is important to engineers, because it derives from an application of previously known scientific knowledge. Magnetic recording was invented by Valdemar Poulsen in 1898. However, the process did not become of general use until the invention of a-c biasing in the 1920s; many subsequent engineering improvements (by many individuals), as well as much additional scientific research, were required before the public could reap the full benefits. Today, of course, magnetic recording is taken for granted and is central to the rapid development of digital computers. But it should be noted that full commercial utilization did not take place until more than 50 years after the original invention.[1] Furthermore, it does not appear recently that the time between invention and commercialization is shortening very much, contrary to popular impression. The heart pacemaker was invented in 1928 but did not reach commercialization until 1960—a time span of 32 years. Electrophotography took 22 years. On the other hand, the video tape recorder only took six years. The principal factor governing the time span between invention and commercialization has been identified as the degree to which the technology supporting the invention has been developed. In the case of the video tape recorder, the underlying technology was available at a high level of development.[2]

The foregoing examples permit us to classify technical creativity in the following ways:

1. *New scientific theories and knowledge.* This is the domain of scientists. Some engineers—those sometimes identified as "engineering scientists"—are engaged in this activity, also.

2. *The "frontier"* where new knowledge is being sifted, explored, expanded, and, in other ways, made usable to society. Here, scientists, engineers, and engineering scientists mix almost indistinguishably; and here, most of the confusion arises as to what is "engineering" and what is "science."

3. *The development of new devices, systems, and structures.* This is the region in which the vast majority of engineers work. Here, the task is to apply scientific, mathematical, economic, and social knowledge to satisfy specific needs. At this point, matters of human usability and of economics become paramount.

---

[1] J. Jewkes, D. Sawers, and R. Stillerman, *The Sources of Invention*, 2nd ed. (New York: W. W. Norton, 1969), pp. 269–272.

[2] *Science, Technology, and Innovation* (Columbus, Ohio: Battelle Columbus Laboratories, February 1973), p. 11.

Within the third category above, many gradations of creative ability are exercised. At the uppermost levels lie such activities as problem recognition and definition, schematic (or "logical") design, and basic *scheme* selection. More will be said about schemes later in the book, but, basically, selection of a scheme means deciding what the method of solution will be, in terms of actual structure. This is the most crucial stage in the life of any embryonic new product and is the phase that will be most critically examined by the Patent Office.

Once a basic scheme has been chosen, there is much additional work to be done in the selection and arrangement of components; these activities involve even the most junior designer, whose only "creative" act may be to decide whether to use five screws or six. All such decisions have their effect: sometimes a whole project fails because of insufficient attention to detail. It is common knowledge that some of our rocket troubles have been caused by "minor" components.

The creative person

Everybody is creative to some degree: creativity is not a step function, so that one is either a thoroughly creative person or else a person completely lacking in this quality. Creativity is distributed by degrees among humanity and is very likely to be closely associated with intelligence.[3]

Psychologists have shown considerable interest in this last point, and some research seems to indicate that there is no correlation between intelligence and creativity. In such cases, however, "intelligence" is likely to mean intelligence *as measured by an I.Q. test* and I.Q. tests measure only certain kinds of intelligence. As Getzels and Jackson put it: "I.Q. tests tend to reward 'convergent thinking,' since they are largely framed in terms of 'acceptable' answers."[4] Since I.Q. tests tend to emphasize such matters as reading comprehension, vocabulary, and recognition-patterns, it can readily be seen that those qualities regarded as "creative" generally are bypassed.

There is some psychological evidence that our educational system may tend to inhibit the development of creativity in individuals. Anne Roe, of Harvard University, suggests there is a "subtle something" about the way in which elementary subjects are taught that may have a stultifying effect upon original thinking. She says, "Teaching these

---

[3] R. B. Cattell, "The Personality and Motivation of the Researcher from Measurements of Contemporaries and from Biography," in *Scientific Creativity: Its Recognition and Development*, C. W. Taylor and F. Barron, Eds. (New York: Wiley, 1963), p. 122.

[4] J. W. Getzels and Philip W. Jackson, *Creativity and Intelligence* (New York: Wiley, 1962).

subjects in terms of 'right' and 'wrong' answers carries a strong moral connotation of considerable significance." Unconventional answers from children may be sweepingly denounced by teachers as "wild" or "silly," but may actually be indications of creative potential. In addition to this, children usually impose sanctions against members of their group who are "different." As a result, potentially creative children often are subjected to severe repressions and frequently develop into isolates.[5]

Some people appear to be natural creators; that is, ideas seem to come forth almost automatically. Others have to engage in deliberate, conscious effort to develop their creative potential. Obviously, there is a limit to each person's ability, but one should take heart from the findings of a California psychologist, that ". . . the actual creative productivity of almost every individual falls far short of his own level of creative capacity."[6]

Clearly, the ideal would be to find ways to induce each individual to utilize more of his natural potential for creativity. Certain techniques have been suggested for this and will be mentioned later, but the first step is to realize that everyone has far more creative capacity than he generally uses. Simple awareness of this can open the door to more effective creative activity. There is no more poetic advice to the would-be creator than that of Ralph Waldo Emerson, who says, in his essay on "Self-Reliance":

> A man should learn to detect and watch for that gleam of light which flashes across his mind from within, more than the lustre of the firmament of bards and sages. Yet he dismisses without notice his thought, because it is his. In every work of genius we recognize our own rejected thoughts; they come back to us with a certain alienated majesty. Great works of art have no more affecting lesson for us than this. They teach us to abide by our spontaneous impression with good-humored inflexibility the most when the whole cry of voices is on the other side. Else tomorrow a stranger will say with masterly good sense precisely what we have thought and felt all the time, and we shall be forced to take with shame our own opinion from another.

Group action: "brainstorming"

For a few years after "brainstorming" was invented, the idea caught fire. It was fun, and marvelous results were claimed. Everybody sat

---

[5] Anne Roe, "Personal Problems and Science," in *Scientific Creativity: Its Recognition and Development, op. cit.*, pp. 133–134.

[6] R. S. Crutchfield, "The Creative Process," in *The Creative Person* (Berkeley: Institute of Personality Assessment and Research, University of California, 1961), p. VI–2.

around the table and gushed forth ideas without restraint. (Indeed, the less restraint exercised, the better.) Then a reaction against brainstorming set in, which was almost as extreme as the enthusiasm had been; heated debate arose as to whether brainstorming was or was not a useful creative tool.[7]

Attempts have been made to establish brainstorming's effectiveness by means of controlled tests; unfortunately, these tests have not been conclusive. Donald W. Taylor of Yale University conducted one such study and decided that brainstorming ". . . *inhibits* creative thinking." In his research, individuals working alone produced more ideas, and more *quality* ideas, than the same number of individuals working as a group under brainstorming conditions.[8]

In another case, two engineers spent more than a month in conceiving 27 possible solutions to a given problem. Then, 11 young engineers having no prior acquaintance with the problem came up with all 27 of these ideas, plus many more, in a 25-minute brainstorming session.[9]

In still a third instance, a pair of psychologists repeated Taylor's tests on different subjects and, this time, found no significant differences between the performance of groups and that of individuals.[10] The major finding of this study was that the key influential factor is the employment of *deferred* judgment, as opposed to *concurrent* judgment, whether used by groups or by individuals. In the exercise of deferred judgment, ideas are generated without any attempt at evaluation until later; under concurrent judgment, ideas are evaluated as they occur.

This, then, is the thing of value that brainstorming can teach us about creativity, on a larger scale: in any such deliberate attempt to be creative, the focus should first be on the generation of a large number of ideas including even some superficially foolish ones. Later, after the flow of ideas has ceased, judgment can be employed while each idea is carefully examined for new leads and hidden possibilities.

Perspiration and inspiration

Thomas Edison is often quoted as having said, "Genius is 99 percent perspiration and 1 percent inspiration." There is another maxim that "inspiration most often strikes those who are hard at work."

---

[7] B. S. Benson, "Let's Toss This Idea Up . . . ," *Fortune*, October 1957, pp. 145–146.

[8] D. W. Taylor, "Environment and Creativity," in *The Creative Person, op. cit.*, p. VIII–5.

[9] E. K. Von Fange, *Professional Creativity* (Englewood Cliffs, N.J.: Prentice-Hall, 1959), p. 51.

[10] S. J. Parnes and A. Meadow, "Development of Individual Creative Talent," in *Scientific Creativity: Its Recognition and Development, op. cit.*, p. 318.

Outstandingly creative persons are almost always noted for their great energy and drive.[11] Nevertheless, many people still believe that inspiration comes unbidden, at idle moments, and can strike only those who are blessed with a mysterious "gift" of some sort. At one time, the U.S. Supreme Court even clothed this notion with official dignity, by insisting that each potential patent be tested to determine if the invention resulted from a "flash of genius." Fortunately, by act of Congress, this idea has been discarded, and inventions are no longer tested on the basis of the manner in which they were made, but on their intrinsic nature.[12] Even so, investigations made by psychologists show that there actually is something called inspiration, although it most certainly does not come unbidden. Instead, it requires a most strenuous preparation period.[13]

This preparation (the "perspiration" part) consists of an intense period of study and search, while one learns everything he possibly can about the subject, followed by a period of extreme concentration and effort during which the "creator" makes repeated attempts to solve the problem. A long-time associate of Edison's told of coming in on the great inventor one night, to find him surrounded by piles of books he had ordered. "He studied them night and day. He ate at his desk and slept in a chair. In six weeks he had gone through the books, written a volume of abstracts, made two thousand experiments . . . and produced a solution." Edison was 24 at the time.[14]

Another example of such diligence is the discovery of vulcanization by Goodyear: the U.S. Commissioner of Patents declared in 1858 that Goodyear had made himself such a master of the subject of rubber that nothing could escape his attention.[15] Thus, the "accidental" discovery of vulcanization was preceded by an intense period of preparation.

W. H. Easton, in "Creative Thinking and How to Develop It," gives this graphic description concerning the occurrence of inspiration (called "illumination"):

> In this case, the thinker encounters a problem of great difficulty; but, as he has no way of knowing this in advance, he proceeds as usual,

---

[11] R. S. Bloom, "Report on Creativity Research by the Examiner's Office of the University of Chicago," in *Scientific Creativity, op. cit.*, pp. 253–258.

[12] *U.S. Code*, title 35, chap. 10, par.. 103. The "Legislative History" of this section states, ". . . it is immaterial whether it [an invention] resulted from long toil and experimentation or from a flash of genius."

[13] B. Ghiselin, "The Creative Process and Its Relation to the Identification of Creative Talent," in *Scientific Creativity, op. cit.*, p. 356.

[14] M. Josephson, *Edison* (New York: McGraw-Hill, 1959), p. 94. Used by permission.

[15] Jewkes et al., *op. cit.*, p. 50.

expecting to clear up the matter without much trouble. This, however, he fails to do. The problem resists all of his initial efforts to solve it, and before long, he discovers he has run into a serious obstacle.

This is a critical point in his work. If he were like most people, he would stop here, giving up the problem as hopeless. But, being a creative thinker, he refuses to accept defeat, so he works on.

But no amount of deliberate thinking gets him anywhere. He develops and applies every promising method of solving his problem he can imagine, but all prove failures.

After struggling for hours, he runs out of ideas. Further thinking is useless, but his intense interest in the matter prevents him from stopping. Yet all he can do is to mill old ideas around in his mind to no purpose. Finally, frustrated and utterly disgusted with himself, he throws the work aside and spends the rest of his day in misery.

Next morning he wakes oppressed. His problem is still on his mind and he thinks about it gloomily. But as the fog of sleep clears from his brain, the tenor of his thoughts changes.

If, now, nothing distracts his attention, he soon finds that exactly those ideas he strove so hard to grasp the day before are now flowing through his mind as smoothly and easily as a stream flows through a level meadow. This is illumination.[16]

Blocks to creativity

Probably the most frustrating block to creativity is what psychologists call "persistence of a misleading set."[17] This means that one solution to a problem is already known and try as one might, one cannot get his mind past that particular solution to see what other (and perhaps better) solution might exist. He says to himself, "all right, here we go for a solution of a totally different sort," but though he strenuously resists it, his mind circles about and comes to rest directly upon the old familiar solution. Successful creative people say they sometimes get around this kind of block by forcing themselves to adopt extravagantly unorthodox viewpoints of their problem, such as, "Suppose I completely inverted this structure and made the output into the input?" or "Now that I have a mechanical solution to this problem, suppose I deliberately try to make one that is completely electronic?"

Closely allied to the block of a misleading set is the one called "functional fixedness." Here, a potential new use for a famiilar object

---

[16] *Creative Engineering, op. cit.*, pp. 6–7.
[17] R. S. Crutchfield in *The Creative Person, op. cit.*, p. VI–8.

is obscured by its present use. An excellent illustration of this kind of block is given by Harold Buhl:

> Some students were once given the task of removing a ping-pong ball from a rusty pipe that had been bolted upright to the floor. In the room with the pipe, students found hammers, pliers, soda straws, strings, pine, and an old bucket of dirty wash water. After fishing vainly with the various tools most of the students finally saw a solution; they poured dirty water into the cylinder and floated the ball to the top. Then the experiment was repeated on other students with one important change; instead of the bucket, there was a crystal pitcher of fresh ice water surrounded by shining goblets on the table with a gleaming white cloth. Not one student solved the problem because no one could connect the beautiful pitcher and its clean water with the rusty pipe.[18]

Premature criticism has caused many an idea to be stillborn. It is better for the potential creator to adopt an attitude of optimistic reserve, always expecting the most favorable results from new avenues of thought. This, of course, is nothing but the application of deferred judgment. Eventually a choice must be made from among the various alternatives, but a possibility should not be eliminated too early by someone's saying, "Oh, that's ridiculous!" For it just might not be so absurd as it seems.

Last, and most pernicious of all, is the block of fear: primarily fear of social disapproval, or perhaps, of supervisorial disapproval. Undoubtedly, fear is at the root of human tendencies to conform. Many writers of the "social protest school" have lashed out at conformity. It should suffice to say that anything really new is a departure from past practice and, therefore, represents some individual's nonconforming. If a person is so afraid of failure as to be unwilling to depart from tradition, he will never be very creative.

A "formula" for creativity

Besides the exhortation to work hard, other specifics may help to stimulate one's creative capacities. To begin with, a simple *awareness* of the different phases[19] of the creative process may be valuable:

1. PREPARATION. It is essential to obtain every possible scrap of knowledge concerning the specific problem that one can. In opposition

---

[18] H. R. Buhl, *Creative Engineering Design* (Ames, Iowa: The Iowa State University Press, 1960), p. 55.

[19] D. W. MacKinnon is source for names of the five phases of creativity. See "The Study of Creativity," in *The Creative Person, op. cit.*, p. I–1.

to this, some people point out that many technical bottlenecks have been broken by men who were novices in the particular fields of their accomplishments. Such successes may be attributable to these individuals' freedom from functional fixedness and from misleading sets. Despite this, most engineers will be more successful at being creative if they first go to the trouble of making themselves *knowledgeable*. The popular supposition that most great inventors of the nineteenth century were lacking in scientific knowledge stems from today's lack of appreciation of the possibilities of self-education.[20] Most of these nineteenth-century inventors were as knowledgeable in their chosen fields as the leading scientists of their day and, in fact, were to a large degree in intimate association with the scientists. Edison, one of the most prolific inventors the world has ever seen, generally avoided *scientists* but made it a rule to gain access to every bit of *scientific knowledge* he possibly could. Whenever he moved into a field with which he was unfamiliar, he first collected all of the published material on the subject that he could and then digested it in an orgy of reading.

2. CONCENTRATION. One way of getting started at being creative is to sit down with the intention of being creative. At first, there may be no discernible result, but each step is a necessary precursor to those which follow. One way or another, one must get himself thinking, long and hard, about possible solutions to his problem, and developing as many promising leads as possible. Concentration is the hardest part of being creative, but it is also the most characteristic.

3. INCUBATION. Incubation is defined as a temporary withdrawal of the conscious mind from the problem while the subconscious continues to work on it. Some psychologists question the necessity for an incubation period. However, it is probably not accidental that the term to "sleep on it" has become common in our language.

4. INSPIRATION. As previously described, inspiration is the sudden appearance of new insight, accompanied by exhilaration and elation. One psychologist points out that the elation is generally accompanied by feelings of certainty, which are not always valid.[21]

5. VERIFICATION. This final period requires steady nerves and enormous determination as one "proves out" his idea, both to stave off despair as obstacles are encountered and to prevent oversights that might result from too much mental intoxication carrying over from phase 4.

---

[20] Jewkes et al., *op. cit.*, p. 64.
[21] B. Ghiselin in *Scientific Creativity, op. cit.*, p. 356.

Deliberate attempts have been made to teach the creative problem-solving process. Parnes and Meadow report that research on the effectiveness of such courses has been conducted at the University of Buffalo, with strongly encouraging results. In the teaching of creative problem-solving, various blocks to creative thinking are first discussed. Some of those cited are difficulty in isolating the problem, rigidity of narrow viewpoints, trouble identifying fundamental attributes, conformity, excessive faith in logic, fear, self-satisfaction, perfectionism, negativism, and reliance on authority.[22]

After the blocks have been identified, the principle of deferred judgment is introduced, together with practice in "attribute listing." In "attribute-listing," the student is taught to look for *fundamental* attributes of an object, rather than to focus on its known functions. For example, in considering a piece of paper, a student might discover potential new applications for paper by studying such fundamental attributes as its whiteness, its square corners, its straight edges, or its translucence. This exercise is expected to help him avoid functional fixedness, an evil that might easily occur if he were to focus prematurely upon the known function of paper as a material for writing.

Students are also taught to keep notes on all ideas that come to them and to allocate definite times for deliberate idea production. They are urged to list all conceivable facts that might relate to their problems, together with lists of questions and possible sources of answers. Potential answer-sources are then followed up. Many people have reported amazement at the number of answers that can be obtained simply by consulting a library. (For some perverse reason, libraries are often among the last sources tapped for answers.)

In the University of Buffalo research on the effectiveness of such courses in creativity, individuals were tested on their creative abilities both before and after taking creative problem-solving courses. The results were then compared with control groups. The findings were that a significant increment in creative ability was produced by taking a problem-solving course; follow-up research showed the effect was a lasting one.

Rewards for creativity

Entrenched in American folklore are stories of lone inventors who became rich. There certainly have been a few such men; there have even been some "Cinderella cases" in which huge financial judgments were granted to inventors after lengthy court battles. However, the

---

[22] S. J. Parnes and A. Meadow in *Scientific Creativity, op. cit.*, pp. 311–320.

chances of someone's being successful in this way are about the same as his getting to be a world-famous movie star or circus acrobat.

A large proportion of inventive people eventually decide to give up such tenuous chances of striking wealth and become corporation inventors. In exchange for their creative talents, they receive a regular salary and fringe benefits, plus a reasonable amount of security. Often the salaries are excellent and may even permit a moderate amount of luxury; this is especially the case when the inventive person displays high-level creativity at regular intervals.

Such arrangements are made because it is virtually impossible for the lone person to underwrite and develop an idea by himself, no matter how good his idea might be. Despite this unavoidable truth, many corporation inventors believe they have been somehow cheated if the company makes big profits on one of their ideas. Such injured feelings stem from a basic philosophical consideration: the idea did not exist until the inventor conceived it and gave it to the world. This is an important thought and one not to be passed off lightly.

Another thought, also to be weighed, is as follows: material progress in this world is completely dependent upon the provision of capital; therefore, the risk capital put up by a corporation is at least as essential to success as the idea itself.

It would be foolish to ask, "Which point of view is right?" For there is no issue of "rightness" here, but only one of scarcity. As long as both capital and inventive talent are scarce, a high premium will be placed on both.

Sometimes the quantity of patents stemming from a given individual is taken as a measure of his creative ability. But patents are not a reliable indicator. For one thing, too many worthless patents have been, and are being, issued. For another, many important kinds of creative activity do not culminate in patents. The correct analysis of a crisis on the production line and the subsequent discernment of its solution can require great creativity (and may save the company as much money as several patentable ideas would earn for it); yet, the result probably will not be patentable.

Some people strongly believe that inventors are legally and morally entitled to share in the profits made from patents, on a royalty basis. Some foreign countries have such plans, but they have not become popular in the United States. Here, many companies pay a token sum (perhaps $50 or $100) to an inventor upon his filing a patent application and a similar amount when the patent is issued. However, the most commonly held belief in this country today is that creative contributions should be compensated in the form of salary.

A 1964 survey investigating the patent practices of 251 U.S. industrial concerns disclosed that three fourths of the companies replying made no cash award whatsoever to their inventors for patentable ideas. Nearly all of the remaining 25 percent had moderate awards of the type just described. A conspicuous exception was IBM, which had installed a plan in 1961 that awards $5000 and more for significant inventions.[23]

Suggestion systems

Most companies have, at one time or another, considered the possibility of installing a suggestion system. The purpose of such systems is to tap hidden wells of creativity in the organization (presumably outside the normally creative departments) and, thus, to gain cost-saving ideas. Sometimes the cash awards from suggestion systems have been of sufficient magnitude to make newspaper headlines, and upon rare occasions, have been truly spectacular.

It is to be wondered if sufficient consideration has been given to the effect such awards may have upon the departments that are in the daily business of improving the company's products and of conceiving, and putting into practice, cost-saving ideas; a few departments of this type are: development engineering, production engineering, manufacturing engineering, and industrial engineering. To a person in such a department, it seems that an outsider can receive an award, amounting perhaps to a year's salary, for an idea that is no better than the ones the "creative engineer" turns out every week. If such a man feels sufficiently outraged, he might even threaten to quit the engineering department and go to work in the shop so that he can be better paid for his engineering ideas.

Coupled with this is the embarrassingly small number of really good ideas acquired through suggestion plans: inevitably, there will be a veritable avalanche of ideas at the inception of one of these plans. After the sifting and sorting, the evaluation committee will generally have nothing to show for its efforts, except a feeling of harassment. Of course, there is always the chance of a really good idea coming along. This chance is what keeps the committee going; meanwhile, the engineering departments, which justify their existence because of their ability to produce workable ideas on a routine basis, feel overlooked, unappreciated, and jilted.

If an evaluation committee is prepared to cope with the frustrations resulting from having to reject 999 suggestions out of 1000 and if the cash awards are not set so high as to bring about internal struggles, then there is nothing inherently wrong with a suggestion system. The rare occasions upon which the committee members discover a good

---

[23] C. G. Baumes, *Patent Counsel in Industry* (New York: National Industrial Conference Board, 1964), p. 35. (One hundred eighty-three companies responded to the question about cash awards.)

idea are sufficient to give them heart and hope, as they brace themselves to face the rest of the pile.

In 1951, a major oil company issued an invitation to inventive Americans to send their ideas to the company's new multimillion dollar research laboratory, where the ideas would be evaluated and the promising ones selected for further development. Three years and thousands of suggestions later, the company concluded it had essentially nothing to show for its efforts. Out of the thousands of ideas submitted, only three had seemed worthy of further testing. Two of these were judged impractical, and the third was turned back to its submitter for development.[24]

---

[24] E. L. Van Deusen, "The Inventor in Eclipse," in *The Mighty Force of Research* (New York: McGraw-Hill, 1956), pp. 74–75. By the editors of *Fortune*.

TWELVE

Design and development

In engineering education circles, an enormous amount of heat has been generated over the word "design." Some educators are alarmed because, in recent years, courses with design content have been largely squeezed out of engineering curriculums by courses in theory and analysis. On the other side, some educators fear that courses labeled "design" are substantially equivalent to the much-despised activity of "handbook engineering" and, therefore, should be abolished.

Meanwhile, within the profession itself, engineers take it for granted that their natural function is *to design* equipment and structures and worry very little about what their activities are called. In truth, some of them may be inclined to call the things they do "research" or, perhaps, "development," but this is probably because these last two words have more prestige value than the term "design."

Even though definitions are usually boring, it seems appropriate to introduce definitions of some key words to avoid misunderstandings:

> **design:** Engineering design is the process of applying the various techniques and scientific principles for the purpose of defining a device, a process or a system in sufficient detail to permit its physical realization. . . . Design may be simple or enormously complex, easy or difficult, mathematical or non-mathematical; it may involve a trivial problem or one of great importance.—Massachusetts Institute of Technology, COMMITTEE ON ENGINEERING DESIGN.[1]

> **development:** . . . technical activity concerned with *nonroutine* problems which are encountered in translating research findings or other general scientific knowledge into products or processes. . . . The engineering activity required to advance the design of a product or a process to the point where it meets specific functional and economic

[1] "Report on Engineering Design," *Journal of Engineering Education*, April 1961, pp. 645–660.

requirements and can be turned over to manufacturing units.—NATIONAL SCIENCE FOUNDATION.[2]

applied research (sometimes called "developmental research or "engineering research"): . . . Investigation directed to discovery of new scientific knowledge and which [has] specific commercial objectives with respect to either products or processes.—NATIONAL SCIENCE FOUNDATION.[3]

basic research (sometimes called "fundamental research"): . . . original investigation for the advancement of scientific knowledge and which [does] not have specific commercial objectives.—NATIONAL SCIENCE FOUNDATION.[4]

Although these definitions have reasonably wide acceptance, they have not been universally accepted. If one does accept them, he can see that the term "development" is included within the meaning of "design," but describes the nonroutine end of the design spectrum. Furthermore, a difficult or novel design project is almost certain to involve applied research; consequently, some might say that "design" embraces applied research, too, although this would seem to be pushing the all-inclusiveness of the word too hard.

Much criticism has been directed at engineering educators, for their apparent failure to prepare their graduates for the design function. For example, an aerospace executive has said:

> . . . the over-emphasis on engineering science is producing people who are as hypnotized by analysis as were the handbook engineers hypnotized by gears, clutches, and bolts. The entire critical area of synthesis is being neglected in favor of an overwhelming emphasis on analysis.[5]

The MIT Committee on Engineeing Design (see footnote 1, this chapter) encountered the following criticism, during its investigations:

> Recent engineering graduates were criticized for unwillingness and inability to consider a complete problem such as a design problem. Instead they showed a desire to seek a fully specified problem which could be answered by analytical methods. It was stated that engineers with advanced degrees were even more prone to avoid a complete

---

[2] *Instructions for Survey of Industrial Research and Development During 1964* (Washington, D.C.: U.S. Dept. of Commerce, 1965), p. 4. Source is the National Science Foundation, for whom the survey was made. Italics have been added by the author.

[3] *Ibid.*

[4] *Ibid.*

[5] W. J. Schimandle, "Science and Engineering in Space," *Seminar Proceedings: Mechanical Design of Spacecraft* (Pasadena: Jet Propulsion Lab., California Institute of Technology, 1962).

> problem. . . . In short, young engineers feel at home in solving problems which have numerical answers—the kind of problem used in school for teaching analytical techniques.[6]

No doubt, the blame for the conditions just expressed belong at the door of the educators, for at least two reasons: 1) because of the frequently held assumption, among educators, that the engineering student can be exposed to four years (or more) of highly structured problems that emphasize analytical techniques and, yet, experience little difficulty after graduation when he discovers that the problems he must henceforth deal with are almost totally unstructured; and 2) because of the recent efforts of educators to cram more and more material into a four-year curriculum. The last circumstance came about substantially as follows: World War II taught engineering faculties that the typical engineer was unable to cope with the problems of the day, because his background in science and mathematics was insufficient. After the war, educators set about the job of getting more science and math into the engineering curriculum but were naturally reluctant to increase its length. As a result, much traditional material was squeezed out (and most of it rightly so). In general, the eliminated material was too specialized, too repetitious, or too elementary and belonged more to the realm of the technician than that of the engineer. However, in the process of thrusting all the desired science and math into the program, many portions of design courses involving creative synthesis (the "good" part of design) were squeezed out, along with those portions that were too specialized, repetitious, or elementary (the "bad" part).

The design process

The great interest in design has stimulated the publication of a number of books that emphasize the design *process*, as distinct from the design of technical hardware in itself.[7] Some of these focus on

---

[6] From "Report on Engineering Design." The MIT Commiteee interviewed leading engineers in the following fields: airplane design; machine tool design; design of bridges, tunnels, and airports; design of diesel engines and gas turbines; design of electrical machinery; operations research; design of electronics systems; design of nuclear submarines; design of chemical plants; design of communications systems.

[7] For example, see:

M. Asimow, *Introduction to Design* (Englewood Cliffs, N.J.: Prentice-Hall, 1962).

D. H. Edel, Jr., *Introduction to Creative Design* (Englewood Cliffs, N.J.: Prentice-Hall, 1967).

J. C. Jones, *Design Methods* (New York: Wiley, 1970).

E. V. Krick, *An Introduction to Engineering and Engineering Design* (New York: Wiley, 1965).

techniques of optimization and have, as one objective, the optimization of the entire process of design, as well as optimization of the components of the design. Analysis of the design process can become very complex and not much detail will be presented here. In fact, only five major phases in design will be identified:

1. Problem definition
2. Invention
3. Analysis
4. Decision
5. Implementation

PROBLEM DEFINITION. Problem definition is one of the steps an inexperienced engineer is likely to skip entirely. Having been subjected to an intensive schooling during which his problems were almost always given to him in clearly defined form, he is uncertain of just what to do, when faced with a problem that is vague and largely unstructured. Since he is thoroughly familiar with *analysis*, he may be inclined to get into this phase as rapidly as possible and may start analyzing something before being completely sure he is analyzing the correct problem. Naturally, there is a real risk that he will wind up with an answer to a problem nobody is interested in. Therefore, the first task of the engineer is to *find out what the problem really is.*

Very often, the original statement of a problem is vague and may even be entirely misleading. Sometimes, merely accepting stated constraints at face value could be the wrong thing to do. For example, a problem statement from the chief engineer might be, "Design an electrical meter that will measure the torque in a rotating shaft without using slip-rings." The constraint involving slip-rings might conceivably have been issued because the chief engineer once tried to use slip-rings on something and found them objectionable for their electrical noise. The constraint, then, actually is against noise, rather than against slip-rings. The engineer working on this problem might discover that the use of improved slip-rings could give him an outstanding design, while limits are maintained on noise. In presenting his arguments to his boss, the engineer must, of course, be tactful in explaining why he failed to follow orders, but the basic point is that the original problem statement should be critically examined to see if it is saying the things that need to be said.

An important aspect of problem definition that is frequently overlooked is *human factors*. Matters of customer use and acceptance are paramount. (It has already been mentioned that two out of three "successful" R & D projects are commercial failures, mostly because of lack of market acceptance.) Consideration also must be given to physical limitations imposed by human capabilities. This is a large subject in itself, and numerous books have treated the topic, especially with reference to such items as average human body dimensions, reach, strength; visual acuity; average manual dexterity; sensitivity to noise, shock, and vibration; tolerance to environmental factors such as temperature, humidity, and acceleration.[8]

INVENTION: THE MAKING OF SCHEMES. Engineering, by definition, is concerned with new things. This is precisely what invention is: the coming up with new ideas. A new idea may involve a combination of old components, but if a new and useful effect results from this combination, then invention has taken place. For the present, any questions of patentability will be excluded from this discussion, and we will concentrate upon the creative act of conceiving an idea for hardware that may solve a particular problem. This potential assemblage of hardware is what was referred to, in an earlier chapter, as a "scheme."

At this stage of affairs, one cannot be sure that the scheme in hand will actually solve the problem. However, the purpose of the next phase of design—analysis—is to shed some light upon the probability of success. Before analysis can begin, though, at least one scheme has to be generated; otherwise, there is nothing to analyze.

As an example of what is meant by a "scheme," suppose that a supervisor instructs an engineer to design an electric accelerometer. The instrument is intended to measure the acceleration of an automobile and is to be mounted on a dashboard for visual reading, although it must be possible also to use the output of the accelerometer to make a record on paper. Hopefully, the engineer will come up with many systems of hardware that, in principle at least, will perform the desired function, but for the purposes of this book, only two of his systems will be described. The engineer's two "schemes" are:

1. Mount a mass to be as free from friction as possible and provide a spring so that, as the automobile accelerates, the inertial resistance of the mass will cause the spring to be compressed. Connect the mass to an iron core mounted within a coil, so that displacement of the mass (and core) will change the inductance of the coil and, thus, give an electrical indication of acceleration.

2. Gear a d-c generator to the drive shaft of the automobile and hook it in series with a capacitor, so that the current in the circuit is proportional to the acceleration (a differentiating circuit).

Our engineer realizes that many questions of feasibility have still been left unsettled in both these schemes, but that is what the analysis phase is for—to answer such questions. The important thing is that he now has something physical and concrete to analyze.

---

[8] *Introduction to Creative Design, op. cit.*, pp. 73–93.

One of the most important aspects of scheme-selection to be noted here is that the engineer has actually posed a textbook type problem for himself. For either of the foregoing examples, he has described a physical structure and has then asked himself, "How does the current vary with car speed?" Nobody gave him the problem in this form. He had to pose the problem for himself, and this posing is one of the most important elements in the design process.[9]

ANALYSIS. It is primarily because of the analysis phase that engineers go to college. All of the other functions—problem definition, invention, decision, and implementation—can be carried out, to a very large extent, without the benefit of a college education. (This helps to account for the remarkable success in design achieved by many non-college men in the past.) Even today, much design can be performed without recourse to the analysis step: the designer proceeds directly from invention to decision, on the strength of experience and intuition.

However, the basic rationale for engineering is that a *better* job of design can be done through the intelligent application of science and mathematics. In fact, some of today's more difficult design tasks can be accomplished only with the assistance of advanced mathematics and scientific know-how. It was once a popular expression that the engineer can do, for one dollar, what the untrained person requires two dollars to do. But an untrained person could not design a jet liner or a satellite even with an unlimited amount of dollars.

The basic analytical tools used by the engineer are mathematics and a collection of scientific "laws." Applying his knowledge of science, the engineer constructs for himself a mathematical model and then, by means of mathematical manipulation, extracts from this model the information he needs. The model should be reasonably representative of the physical system and, obviously, should be no more complex than is absolutely necessary to produce the required information. Herein lie two common errors of inexperienced engineers: 1) frequently, the model chosen only slightly represents the physical system but has been chosen primarily because it is one the engineer knows how to analyze or because it is elegant; 2) far too often, the model is more complex than it has to be; a lengthy and involved analysis could be a complete waste, for example, if the purpose of the analysis is to produce order-of-magnitude figures for a comparison with competing schemes.

If the engineer possesses insufficient scientific information to construct a good mathematical model, he may have to initiate a research program to get the information he needs. Therefore, the analysis phase

---

[9] See D. W. Ver Planck and B. R. Teare, Jr., *Engineering Analysis, An Introduction to Professional Method* (New York: Wiley, 1954).

of design may include research. In fact, *all* of the first three phases of design may require research before a man can properly define a problem; test out a scheme and, thus, arrive at an invention; or obtain the scientific information necessary for the purposes of analysis.

The most useful thing about the analysis phase of design is its production of quantitative information that can be used as a basis for scheme selection, as opposed to the purely qualitative nature of the invention phase. Thus, it is necessary to get *numbers*, at this point, and to become specific about the hardware nature of components and their interconnection. Through mathematical analysis, the influential system parameters can be identified and optimum values selected. After this has been done, at least in a preliminary way, the decision phase can follow.

DECISION. Even after mathematical analysis, individual judgment is necessary. For one thing, economic considerations enter the picture at this stage. The product must be produced at a low enough cost, and at sufficient volume, to recover all of the development expenses and produce a profit.

Selection of a particular scheme will depend upon which of the various schemes offered appears the most favorable, as a result of optimization. Optimization is finding the best combination of certain variables that will maximize a desired result. This desired result is known as a "criterion function." But in establishing the relative importance of various criterion functions, we are forced to resort to value judgments. Through the application of value judgments, the engineer can state how much the different aspects of the design are worth to him, that is, the appearance, the weight, the durability, the selling price, the serviceability, the quietness, the sensitivity, and so on. If all these can be placed together on a value scale, it is possible to pick out an optimum combination and to make a single selection from among the competing alternatives. Optimization is too extensive a subject to be treated here, beyond the general indication of its nature that has just been given.[10]

IMPLEMENTATION. Before the design process may properly be considered complete, detailed manufacturing instructions must be prepared so that the device, structure, or system can be produced. Historically, the medium for such instructions is the working manufacturing drawing, although computer output data are beginning to be used, in the form of punched or magnetic tape.

However, even before working drawings (or tape) can be produced, much detail design, involving spatial considerations, strength, weight,

---

[10] For more information on optimization, see M. Asimow, *Introduction to Design, op. cit.*

economy of manufacture, and the like, is necessary. This detail design is customarily carried out on the layout board, and to many people, this is what the term "design" means. If one adds to this the fact that many nonprofessionals, who are given the job title of "designer," engage in this activity, it can be seen why so many professional engineers are emotionally biased against the word "design."

Although detail design is only one of the many phases of design it is an important phase and one that cannot be abandoned by the professional engineer. His responsibility for a design extends to the last detail, though he will have many technicians to assist him in the latter stages of this endeavor. One of the things to be hoped for in the future is that large numbers of qualified people will take advantage of the two-, three-, and four-year educational programs that specifically aim at this phase of design.

There used to be a fear, on the part of many new engineering graduates, that they might get "stuck on the drawing board," but today, this is unlikely to happen in most companies. Engineers are paid high salaries because of the special analytic tools they possess, and it is wasteful to use them as draftsmen. Nevertheless, engineers (especially mechanical engineers) should realize that the drawing board can be a very useful tool to them, at times. For example, before it is possible to perform an analysis, it may be necessary to see just how big the parts may be or what spatial constraints may exist on their interconnections. Usually this can be established only by means of a layout on the drawing board; trying to accomplish it through a draftsman can often be inefficient (or even impossible) if the item is extremely complex, simply because of the inadequacy of human language as a communication means.

ITERATION. Lest it be thought that a design project should be expected to proceed neatly down through all the phases directly in order, special mention must be made of *iteration*. What this elegant word means is that the designer frequently may find it necessary to back up and do something all over again. New data may be uncovered, a new idea may be generated, mistakes may be found, or things simply may not work as expected. The last point, *things may not work as expected*, is especially important: careful analysis must be followed by careful testing. Failures on the test bench may even require complete abandonment of a given scheme.

It has been aptly pointed out by A. J. Winter, that *iteration* is one of the special characteristics that distinguishes the design process from analysis or from research. In analysis, the *starting* point is known, and one works through the process until an end point is reached. In design, the *end* point is specified, and a starting point must be assumed. During the process of design, one attempts to work from the assumed starting point to the required end point. As errors in direction occur, backing-up periodically takes place, and this is "iteration."

Undermathematizing and overmathematizing

As has been previously mentioned, a primary reason for requiring engineers to have a college education is to provide them with the tools of analysis—and mathematics is the most powerful of these tools. Nevertheless, two types of engineering graduates with serious misconceptions about the role of mathematics in engineering regularly emerge from school. The first is the *under*mathematizer, who cannot wait to get out of school so that he can forget about calculus and differential equations. The other is the *over*mathematizer who is so deeply impressed by the power of mathematics in physical problems that he thinks mathematics *is* physics. The existence of the undermathematizer is easily accounted for: he has been talking to practicing engineers in industry who have assured him that calculus is useless—after all, *they* never use it. The creation of the overmathematizer is a more subtle process and comes about through basic misconceptions concerning the nature of mathematics.

Exactly what *is* mathematics? It is probably safe to say that most engineers have never asked themselves this question, nor have they ever thought that such a question could be relevant to their activities. Yet the question is highly relevant. The eminent engineer T. von Karman once wrote, "Sometimes we have the feeling with mathematics that we have learned to start the mechanism of mathematical operations, but after the gears begin to work the machine gets out of hand and we do not know what it is doing or where it is going."[11]

It is not an idle exercise to inquire into the nature of mathematics. Far too often, the impression left upon the new graduate is that mathematics controls physical reality—that, moreover, it *is* reality. Without realizing it, such a graduate is adopting a view that is hundreds of years old and was most strongly focused in the philosophy of Immanuel Kant (1724–1804). Kant believed that all propositions could be divided into two types: *empirical* (those that depend upon perception by the senses) and *a priori* (those that have a fundamental validity of their own).[12] According to Kant, mathematics belongs to the latter category, wherein intuitional perception of space and time constitutes an *a priori* frame into which all physical experiences can be fitted.[13]

Two physicists, H. R. Lemon and Michael Ference, comment as follows on the grip that ideas like the foregoing still have upon students:

---

[11] T. von Karman, "Some Remarks on Mathematics from the Engineer's Viewpoint," *Mechanical Engineering*, April 1940, pp. 308–310.

[12] J.F. Morse (Ed.), *Funk and Wagnalls Standard Reference Encyclopedia* (New York: Standard Reference Works Publ. Co., 1963), p. 5341.

[13] H. Hahn, "The Crisis in Intuition," in *The World of Mathematics*, J.R. Newman (Ed.), (New York: Simon and Schuster, 1956), p. 1956.

> Too often our technical students at the beginning of their career are left with the impression that phenomena follow and conform to certain 'laws' derivable from *a priori* grounds and of the utmost mystery as to origin. The fact that the laws themselves—at least the more fundamental ones—have no *a priori* basis but simply describe, generalize, and integrate a great range of interrelated experimental facts too frequently dawns upon the more mature physicist rather late in his own development.[14]

It has been shown that even great triumphs of reason, such as the development of Einstein's theories of relativity, had their genesis in experimental data. In the words of another physicist, George Gamow:

> . . . the abandonment of classical ideas of space and time and their unification in a single four-dimensional picture were dictated not by any purely esthetic desire on the part of Einstein . . . but by stubborn facts that emerged constantly from experimental research, and that just wouldn't fit into the classical picture. . . .[15]

The fundamental point is that mathematics is an *invention* of man; its purpose is to systematize the processes of logic. If the human mind were equal to the task, it could solve problems without recourse to mathematics. For example, it is conceivable that almost any arithmetic problem could be solved entirely in one's head, but most people reach answers faster, and with greater accuracy, by employing certain manipulative rules. Great faith is placed in the validity of the results, but people tend to forget (if they ever knew) that arithmetical manipulations are based upon some very fundamental axioms, known as Peano's Axiom System.

At a more sophisticated level, a system of differential equations will represent a statement of the relationships among a number of physical variables. Once the equations are written down, the answer has already been fixed and is implicitly locked in the equations. We would be able to perceive the answer directly, if the human mind could hold all the variables and their relationships in view at once and could comprehend the simultaneous effect, upon all the dependent variables, caused by changes in the independent variables. The human mind cannot do this, of course, but the desired result can be achieved by the operations of mathematics. In mathematical operations, certain theorems whose validity has been thoroughly established are employed, and if these theorems are correctly applied, then we have confidence that the result correctly represents the solution that was embodied in the original

---

[14] H. R. Lemon and M. Ference, Jr., *Analytical Experimental Physics* (Chicago: The University of Chicago Press, 1943), p. v.

[15] G. Gamow, *One, Two, Three . . . Infinity* (New York: New American Library, 1960), p. 93. Used by permission.

equations. This leads to the final point: the solution can convey no more factual information than was implied by the original formulation of the problem. In other words, mathematics can only transform information into more useful forms; it cannot, of itself, create information.

Mathematics is said to be *analytic*: the truth of analytic statements is self-contained in the axioms with which they begin, in precise definitions and in the logic of the proofs attached to the theorems that are employed. Analytical statements convey no factual information. Factual information is conveyed only by *synthetic* statements, which come from actual experience. The following is a comment by one more mathematician, this time Richard von Mises:

> Occasionally one finds also mathematicians who are of the opinion that physics is reducible to mathematics; they hold, for instance, that electrodynamics has become a 'part of geometry' through the theory of relativity. Such utterances are logical misconceptions and go ill with the critical subtlety which the mathematician otherwise often exhibits.
>
> By the mere manipulation of signs [symbols] according to chosen rules one can indeed learn nothing about the external world. All the knowledge we gain through mathematics about reality depends upon the fact that the signs as well as the rules of transformation are in some wise made to correspond to certain observable phenomena.[16]

The overmathematizer is unduly impressed by the role of mathematics in engineering and believes he is not doing engineering work unless he is doing mathematics. As a result of his point of view, he is likely to overemphasize the analysis phase of design to the virtual exclusion of the other phases. But he is at least in a better position than the engineer who undermathematizes. The overmathematizer can broaden his effectiveness merely by opening his eyes to the existence and importance of the other phases in the design spectrum. The undermathematizer, on the other hand, quickly finds that he has permanently limited himself in the scope of the kinds of problems he can handle. There are simply too many engineering problems in today's world that are too complex for the naked human mind to handle, unassisted by the powerful logical formalism that mathematics can provide.

Industrial design

Normally, industrial design is considered to be outside the sphere of the professional engineer, although it may affect his activities in various important ways.

---

[16] R. von Mises, *Positivism, A Study in Human Understanding* (Cambridge, Mass.: Harvard University Press, 1951).

Industrial design should not be thought of as synonymous with *styling*, although it consisted almost entirely of styling when it began, a generation ago. In the 1930s, an industrial designer was often conceived of as a "wizard of gloss," who could make his clients' sales zoom simply by glamorizing their products. Today, however, most industrial designers have gone in for "total service": not only do they wish to take in hand every aspect of the product that affects the customer—usefulness, safety, ease of handling, styling—but they also would like to plan the client's future product line for him, make over his corporate offices, and refashion the company's "image" through its packaging, trademarks, and letterheads. *Fortune* quotes the the president of one corporation who was talking about the industrial designer his company had retained: "Mind you, I like the fellow. But I sometimes get the feeling that he isn't satisfied to be a designer—he wants to be my right hand and maybe even me."[17]

Some companies retain outside consultants to perform their industrial design, while others have developed complete in-house design establishments. General Motors is one of the latter and employs 1400 people on its styling staff. Styling is very important to GM, and they have even established the office of vice-president of styling.[18]

Matters do not always flow smoothly between engineers and industrial designers, or between engineers and architects (who strongly resemble industrial designers in their viewpoint and in the kind of service they offer). Industrial designers and architects believe that engineers tend to overemphasive economy and technical function and that they give scant attention to relationships between function and appearance, or to esthetics. They could be right. Some of the products of engineers possess a natural beauty all their own; jet aircraft for example, or the Golden Gate Bridge, or cities lit up after dark (provided they are viewed from a sufficiently great distance). However, other products of engineers are natural horrors; power plants, for instance, and many freeways and bridges. In 1964, a storm of protest arose in Portland, Oregon, over the appearance of a bridge that had been designed by engineers of the Oregon Highway Department. It was charged that the bridge had a "mediocre Erector Set" character. The protests were so strong that the city government requested that five alternate designs be submitted on the next bridge to be built, for prior examination by the city's art and planning groups.[19]

In San Francisco, there have been rumblings over the design of the Bay Area Rapid Transit System (BART), the first completely new urban

---

[17] S. Freedgood, "Odd Business, This Industrial Design," *Fortune*, February 1959, pp. 131–132.

[18] A. P. Sloan, Jr., *My Years With General Motors* (New York: Macfadden, 1965), p. 277.

[19] *Engrg. News-Record*, October 8, 1964, p. 56.

transportation system to be designed in the United States in a long time. The *San Francisco Chronicle* editorialized:

> The problem is that the architects are really in the employ of the engineers (instead of the other way around), and engineers are primarily interested in building structures, only secondarily concerned with esthetic design.[20]

Even though not all engineers will agree with the *Chronicle's* views, it is true that engineers must be concerned about esthetics. The public is becoming increasingly interested in beauty in man's environment and is demonstrating a willingness to pay the price to achieve it. It is rare that a competent engineer is also competent in esthetics. As a result, engineers will find themselves permanently in partnership with industrial designers and architects.

Will computers replace engineers?

To the extent that engineers are used for routine tasks, they will probably be replaced by computers. Jobs requiring a great deal of calculating or detail design tasks that can be systematized, are prime candidates for computerization. But no one has yet discovered how to make a computer be creative, although there is plenty of effort in that direction. Computers have already been used to compose music of a sort and even to write low-grade poems. It has been predicted that machines may eventually take over the task of hammering out Tin Pan Alley tunes.[21] However, "creativity" of this type is dependent upon the ability of the computer's designers to specify the creative rules by which the machine operates. Some people believe creativity is the ability to make random connections that turn out to be meaningful. If their view is correct, then computers might eventually be able to perform functions of this type and thus be creative. Nevertheless, the building of a computer with as many random connection possibilities as the human brain has, still lies an unforeseeable distance in the future.

In 1957, some experts were predicting that, within 10 years, a computer would discover an important new mathematical theorem, write music of value, and reduce theories of psychology to computer routines. Long after the 10-year period was over, these things had not happened—to the author's knowledge—although it might be hasty to declare that they will never come to pass.

---

[20] *San Francisco Chronicle*, "This World" sec., April 18, 1965, p. 20ff.

[21] G. Burck, "Will the Computer Outwit Man?" *Fortune*, October 1964, p. 120ff.

THIRTEEN

Patents

The distinguished judge Learned Hand is reported to have said, "I am very little certain about anything, but least of all about patents." Since the judge is one of the nation's outstanding authorities on patent law, it seems highly likely that there is more on the subject than appears at a first reading. He meant that there is scarcely any subject more complex than patents—or more controversial. In spite of this, most Americans reach adulthood with clearly developed, and usually favorable, attitudes toward patents.

Most informed observers agree that a large percentage of issued patents have no commercial value.[1] Another upsetting factor is that well over half of these patents which reach litigation are declared invalid.[2] A study of one five-year period showed that, in 89 percent of all patent cases brought to the courts of appeals and the U.S. Supreme Court, the judgments held the patents to be either invalid or not infringed.[3]

So why should we be concerned about patents at all? For one reason, close to 50,000 patents are issued each year in the United States, and even if only 10 percent of these have value, the total is still an impressive 5000 *valuable* new patents a year. Once a company begins to collect a strong patent position (especially if it includes some court-tested patents), other companies will tend to steer clear and look for greener pastures. For example, the Polaroid Corporation established an impregnable patent structure, at least until some of its basic patents began to run out. Largely as a result of the sheltered position offered

---

[1] J. Jewkes, D. Sawers, and R. Stillerman, *The Sources of Invention*, 2nd ed. (New York: W. W. Norton, 1969), pp. 88–91.

[2] J. C. Stedman, "The U. S. Patent System and Its Current Problems," *Texas Law Review*, March 1964, p. 464.

[3] D. G. Cullen, "Patents in Litigation, 1941–45," *Journal of the Patent Office Society*, December 1946, pp. 903–904.

by its patents, Polaroid's annual sales increased by a factor of 14 in the 1950s.[4]

Unfortunately, many patents have only marginal value, and some even skirt the borderline of trivia. The U. S. Supreme Court has spoken harshly of the ". . . list of incredible patents which the Patent Office has spawned"[5] and has struck hard at the patenting of gadgets. The court has said, "It is not enough that an article is new and useful. The Constitution never sanctioned the patenting of gadgets. Patents serve a higher end—the advancement of science." Some people, however, believe the Supreme Court has gone too far in its high number of patent invalidations. One of these was Supreme Court Justice Jackson, who said acidly, in a 1949 dissenting opinion,

> It would not be difficult to cite many instances of patents that have been granted, improperly I think, and without adequate tests of invention by the Patent Office. But I doubt that the remedy for such Patent office passion for granting patents is an equally strong passion in this Court for striking them down so that the only patent that is valid is one which this Court has not been able to get its hands on.[6]

It is almost impossible to find a restrained, dispassionate statement about patents. Those who defend the practice often take the position that America's great technological progress can be directly attributed to the influence of her patent system. They point, with alarm, to the anti-patent movement that seems to have developed in the United States during recent years.[7] Many supporters of the patent system declare that industrial firms will not undertake expensive development programs unless they can be sure the results can be protected by patents. Executives of some corporations have declared that they would be driven to cut their research and development expenditures drastically, if the patent system were to be abandoned.[8] Yet, others have stated that their innovative activities would not change at all, if there suddenly were no patent system.[9]

---

[4] D. L. Brown, "Protection Through Patents: The Polaroid Story," *Journal of the Patent Office Society*, July 1960, pp. 439–455.

[5] *Great Atlantic & Pacific Tea Co. v. Supermarket Equipment Corp.*, 340 U. S. 147–158 (1950).

[6] *Jungerson v. Ostby & Barton Co.*, 335 U. S. 560–572 (1949).

[7] R. Spencer, "Thinking Ahead: Threat to Our Patent System," *Harvard Business Review*, May–June 1956, p. 21ff.

[8] F. Machlup, "Patents and Inventive Effort," *Science*, May 12, 1961, p. 1463. Dr. Machlup's article essentially declares that "the evidence is insufficient to prove or disprove the claim that patent protection promotes inventive effort."

[9] F. L. Vaughn, *The United States Patent System* (Norman: University of Oklahoma Press, 1956), p. 12.

Practically no one can be found among the critics of the patent system who recommends such an extreme course of action as its outright abolition. One possible exception in the past was Thomas Edison, who wrote with some bitterness during long drawn-out litigation over his electric light patents, "Say, I have lost all faith in patents, judges and everything else relating to patents. Don't care if the whole system was squelched."[10] Edison's bitterness is understandable. Although he ultimately won his electric light suit, the victory cost his company $2 million and is said to have made his adversary, Westinghouse, almost insolvent.

The occurrence of such calamitous litigation is, at the same time, one of the patent system's chief evils and one of its chief advantages. It is bad because of the social waste, advantageous because it makes potential infringers wary of the patents held by others. A prospective litigant would be very foolish to set about breaking another man's patent through the means of costly court actions, unless he were virtually certain, in advance, of the outcome. Herein lies a partial explanation of the high percentage of patents held invalid by the courts: generally, only weak patents are brought to court, in the first place.

The cost of monopoly

"Monopoly" is a bad word in America, and rightly so. It implies outrageous profits and social irresponsibility. Even defenders of the system admit the patent is a form of monopoly, but they point out that it is a *limited* monopoly, given to the inventor in return for his making known something that did not exist before. They hold that the granting of a limited monopoly to an inventor is an incentive necessary to increase the flow of inventions and thereby benefit society.

Nevertheless, critics of the patent system generally base their criticisms on monopolistic grounds. Most often the critics are economists, some of whom hold that the provision of a financial incentive to the inventor may come at too high a cost to the rest of society. A dramatization of this point, attributable to the Princeton economist Fritz Machlup, follows:

> Assume that twenty corporations are engaged in selling a particular product and that 100 million units are sold annually at $1 each. One firm patents an improvement on the product that results in a ten-cent saving on each unit. If all firms were free to use the improvement, the public would be able to acquire the product at ninety cents each; obviously this would benefit the public. However, free use is im-

---

[10] M. Josephson, *Edison* (New York: McGraw-Hill, 1959), p. 355. Used by permission.

> possible, because of the patent. Instead, the product continues to sell for $1, and the innovating firm makes the extra profit, which represents its incentive. If the firm continues to corner one-twentieth of the total market for seventeen years (the life of the patent), it will realize $8,500,000 extra profit, which is presumably enough to cover its research costs and give a return, besides. But during the same period, the public has been required to forgo the opportunity to purchase the 100 million units per year at a reduction of ten cents each: The total cost to society, in lost savings, is $170 million.[11]

Admittedly, the example is oversimplified and extreme. In actual practice, the other firms would also introduce innovations that, in time, would tend to drive the price down. In addition, there is no valid reason to believe the price would remain at 1 dollar, even in the absence of other innovations. The original firm would probably try to increase its share of the market by lowering its price, and this would force the other firms to do the same thing. Still, the example displays in a graphic manner a principal ground upon which many economists base their recommendations for revision of the patent system. Some believe the seventeen-year period is too long; they suggest that ten years, for example, might be sufficient to reap the rewards. Defenders of the system—usually patent lawyers and officers of companies with strong patent positions—lash back at such proposals as disastrous to the cause of innovation and social progress.

Of a somewhat less theoretical nature is the fact that some firms have acquired incredibly large patent holdings: this has caused them to come under government scrutiny as being in restraint of trade. The American Telephone and Telegraph Company (AT & T) and International Business Machines Corporation (IBM) are two such firms. These (among others) have been required to open their patents to licensing at "reasonable" rates. In the case of AT & T, almost 9000 patents were involved.[12]

To summarize, the criticisms of the American patent system are as follows:

1. The rewards to the innovator may be coupled with a disproportionate cost to society.
2. Too many patents of doubtful validity are issued; these impose costly burdens on the nation's judicial system and needlessly impede competition in the American economy.
3. Patent procedures are too slow, although the average pendency time has been reduced from four years (in 1964) to 25 months (in

---

[11] *The Role of Patents in Research—Part II* (Washington, D.C.: 1962), pp. 198–200. (Proceedings of a symposium sponsored by the National Academy of Sciences, National Research Council.)

[12] *Spencer, op. cit.*

1972); the backlog of patents pending in 1972 was 181,259, down from the 1964 high of 219,691.[13]

4. The patent system has lost standing with many inventors, because the illusory rewards it has held forth did not materialize.

The value of patents

In spite of all the criticisms leveled against the patent system, the net judgment of investigators appears to be that the system substantially fulfills the purposes for which it was intended: the stimulation of invention and dissemination of knowledge.

In a sincere effort to get to the heart of the subject, the National Academy of Sciences (NAS) in 1960 appointed a distinguished committee to examine "The Role of Patents in Research." The committee held a symposium, to which they invited some of the people best-informed on patent matters in the United States; the purpose of the symposium was to give full play to all sides of the question.[14] The committee's conclusions were substantially as follows:

1. The patent system does stimulate research and development.

2. The sheltered competitive position offered by patents is especially beneficial to small companies in diminishing the advantages possessed by large companies because of their size.

3. One of the greatest benefits of the patent system is its encouragement of the publication and dissemination of information. Knowing that it is protected by a patent application on file, a company is thereby made willing to release information concerning its developments. The alternative would be industrial secrecy, which would be deleterious to progress.[15]

Some other important advantages of the system (not emphasized in the NAS study) are the following *defensive* uses of patents:[16]

4. Sometimes patents are taken out, not so much to prevent others from entering the field, as to make sure the corporation's freedom in its own field is preserved.

---

[13] *Commissioner of Patents Annual Report—Fiscal Year 1972* (Washington, D.C.: U.S. Dept. of Commerce/Patent Office, 1972).

[14] The committee members quickly discovered that they were investigating, not the effects of the patent system on *research*, but its effects on *development* (in other words, on engineering). In fact, they decided that the patent system and research are scarcely relevant.

[15] *The Role of Patents in Research—Part I* (Washington, D.C.: 1962). (The Committee Report of the NAS, National Research Council.)

[16] To borrow John Stedman's words, patents can be a means of defense ". . . to protect one *against* the patent system, rather than to make use of it."—from J. C. Stedman, "The U.S. Patent System and Its Current Problems."

5. A storehouse full of patents can be trading material that enables a company to cross-license with others and, in this way, obtain rights to adversely held patents.

During the NAS symposium, it was proposed that the American patent system be changed to:

1. Provide for "petty patents" (petit patents) with a life of three to five years for minor inventions, as is now done in some European countries.
2. Provide for compulsory licensing at "reasonable" royalties.
3. Provide for renewal fees which must be regularly paid in order to keep the patent alive.
4. Establish consistent standards of patentability which would be adhered to both by the Patent Office and by the courts.

What is invention?

The mere conception of an idea is insufficient to constitute invention: the invention must be "reduced to practice." Ordinarily, this means that an actual working model of the invention has been constructed and has successfully functioned. (Very seldom does the Patent Office require that an actual working model be sent to Washington to support the application.) However, the simple act of filing an application that discloses an operable patentable structure is called *constructive reduction to practice* and will usually be accepted by the Patent Office. The patentee should realize that if such a patent later becomes involved in litigation, it will be vulnerable to attack; a patent based upon actual working models is much stronger.

As was mentioned before, the term "invention" is extremely difficult to define, and this very difficulty has led to the strife between the U.S. Supreme Court and the Patent Office. Dictionary definitions are useless; one can never be certain his invention is truly patentable until his patent has been judged valid by the Supreme Court. Nevertheless, some general rules for patentability are used by the Patent Office. These are listed briefly in this section. (It should be borne in mind that a treatment of this extent cannot hope to reveal all the complexities of patent matters. The purpose here is to impart an awareness to the engineer that the patent system is like an iceberg, with 90 percent of its substance hidden beneath the surface.)[17]

---

[17] For a more detailed treatment, see R. A. Buckles, *Ideas, Inventions, and Patents* (New York: Wiley, 1957); or A. K. Berle and L. S. de Camp, *Inventions and Their Management*, 3d ed. (Scranton, Pa.: International Textbook, 1951).

Besides the requirement that the invention be new, several rules of invention are commonly applied:

1. A device is not invention if it is obvious to one "skilled in the art." Naturally, a big source of contention is that something which appears obvious to hindsight may not have been so obvious in advance. Some of the best inventions appear to be the simplest in the light of the improvement in vision afforded by hindsight.
2. A combination of old elements is not invention if no new result is forthcoming; such a combination is called an *aggregation*. However, if there *is* a new and unexpected result because of the combination, then the combination constitutes patentable invention. The vast majority of patents that are issued are actually combinations of old elements.
3. A mere substitution of obvious equivalents, or a substitution of materials, is not invention, unless some surprising result is thereby obtained. Also, a mere change in size does not constitute invention, unless there is a new result. A famous example is that of Edison's electric light. Others had built operable incandescent lamps before Edison, and it was claimed that the only difference in his lamp was a reduction in size of the carbon filament. However, Edison's patent was upheld on the basis that it was this reduction in size which caused his lamp to be practical, all previous models having burned for only short periods of time before going out.[18]

A preponderance of evidence

From the engineer's personal viewpoint, one of the most important things he can do to protect his employer's interests is to keep complete, accurate records. For a variety of reasons, it may later be necessary to establish in court the earliest date upon which conception of an idea took place. Further, it may be necessary to prove that the engineer's organization exercised diligence in pursuing an idea. A patent may easily be lost, unless diligence can be proved. In such cases,

---

[18] A highly intersting account is given by Matthew Josephson in "The Invention of the Electric Light," *Scientific American*, November 1959, pp. 99–114, and, also, in his book, *Edison*. Josephson points out that Edison's true genius has been obscured by too much emphasis upon his invention of the electric light. Edison's contribution was of even greater significance: he invented an entire system of electric lighting. Earlier systems required tremendous amounts of copper, since they used low voltages and high currents. Edison conceived the high-resistance, high-voltage (100 volts), low-current system and then proceeded to develop the generator and lamp that made the system practical.

the engineer's notebook almost invariably constitutes the legal evidence that must be produced in court.

Engineers frequently become restless under the procedures laid down for the keeping of notebooks. Such procedural exhortations will generally include the following:

1. Keep a bound, stitched notebook, so that future jurors may be convinced that sheets have not been removed or added. Number all pages.
2. Keep all records in an absolutely uniform manner. Use every page. No not erase, but mark out with neat lines instead. Use ink or indelible pencil.
3. Date and sign every page, and have every page witnessed and signed.

Engineers generally regard this kind of thing (especially the last point) as red tape. However, lawyers know that faithful adherence to these procedures will add up to the preponderance of evidence needed to win cases.

Generally, notebooks will be regularly reviewed by a corporate patent department in order to ferret out overlooked inventions. If an invention has been made, the engineer will be asked to prepare a disclosure. A disclosure describes the invention in meticulous detail and points out its advantages and its potentially patentable features. In addition to the disclosure's possible future use as legal evidence, it will form the foundation upon which any patent application will be based.

After a disclosure has been written, its contents should be described to someone competent in the field (not a man's co-inventor, nor his wife), and then signed by the competent person, as a witness, thus: "Described to and understood by me: (*signed*) (*dated*)." It is better to have two witnesses. The witness(es) should sign, or at least initial, every page. After signing, witnessing, and sometimes notarizing, some companies even go to the extent of inserting a copy of each disclosure into an envelope and mailing it to themselves, via registered mail. These copies are then retained, unopened, against the day they may be needed to prove the date of conception.

In addition to the date of conception, it may be necessary to prove the date upon which reduction to practice took place. Therefore, continuing records of the work must be maintained, especially photographic evidence of model construction and operation, properly signed and witnessed. Such records will also be useful in proving diligence. If any substantial gaps appear in the record, this may be sufficient to constitute abandonment, and the patent may go to a more diligent person even though he has a later conception date.

Some additional items that must be borne in mind by patent applicants are:

1. The true inventor(s) must be identified. Affixing any person's name (such as the inventor's superior, for example) who is not legally and truly the inventor or a co-inventor is illegal.
2. The invention cannot have been patented or publicly described before.
3. If the invention is placed on public use or sale or is described in a printed publication, the inventor must file his application within one year or he will forfeit his rights to the patent.
4. Laws of nature, systems of logic, methods of doing business, mental processes, theories, and systems of mathematics are not patentable.
5. There are three basic kinds of patents:
   a. *Utility patents*, which may be granted on processes, machines, manufactured articles, or compositions of matter, are good for seventeen years.
   b. *Plant patents,* which protect the developers of such things as special kinds of roses.
   c. *Design patents*, which cover external appearance features and may be taken out for 3½, 7, or 12 years at the applicant's option, the only difference being the size of the fee.
6. Patents are issued to individuals, not to corporations. However, the inventor may *assign* his patent to a corporation. In many cases, the assignment is simultaneous with the filing of the application.

Anatomy of a patent

Patents are very imposing-looking documents. Many run to scores of pages, some to hundreds. Characteristically, there will be page upon page of complicated drawings followed by pages of even more complicated text and, finally, at the end, the most complicated part of all: the claims. There is a natural tendency to assume that the entire structure disclosed is covered by the patent. Such is seldom the case, however; the only part of the patent that has any legal force is that which is specifically contained in the claims.

First come the *illustrations*, with every part identified by a key numeral that will be referred to in the descriptive material. Shading, cross-hatching, and the use of line weights are all rigidly prescribed.

The *specification* is a description of the device, complete as to every detail, so that a competent workman could construct an operable model from this part of the document. The beginning of the specification generally presents a statement of the objectives plus a brief resumé showing how the invention achieves these objectives and how it is an improvement over the prior art. Much "art" is taught by the description, and from the viewpoint of the dissemination of knowledge, this is the most important part of the patent.

The *claims* form the only part of the patent that gives any legal protection to the inventor. Even though a patentable structure may be described in the specification, it can be seized upon by competing companies unless it has been described by at least one of the claims.

The claims are usually very upsetting to the newcomer to patents because, often, they all sound very much alike, extending on and on into tedium. There are excellent legal reasons for writing them this way, however, and the future value of the patent is going to depend upon the skill, knowledge, and ingenuity of the claims-writer. Even for a basic invention, much of the potential protection can be lost if the claims are poorly drawn.

Generally speaking, the first claim will describe the invention in the broadest possible way, and the succeeding ones will become progressively more specific. It should be borne in mind that, the more specific a claim is, the narrower it is and the probability increases that someone will be able to get around it. As just one example, if a claim were to use the words "a straight member," and if a curved one would do as well, then the use of a curved member would avoid the claim.

Another factor in favor of multiple claims is as follows: if a patent has only one claim and that claim is ever invalidated in the future, the whole patent is gone. But if there are many claims, graded in their degree of specificity, it is possible that adverse judicial action may not invalidate all of them. In this way, some of the coverage would still be intact. Many patents have large numbers of claims (frequently as many as 40 or 50). The all-time record-holder was issued in 1915 to Gubelman and had 797 claims. It had been pending for 26 years.[19]

Even with careful reading of a patent, it is virtually impossible to tell, with any precision, just what coverage it really offers. It is necessary to review the prior art (previously issued patents) to discover exactly what significance the claims possess. It is also necessary to obtain a copy of the "file wrapper" of the patent, which contains the complete record of all the transactions between the inventor's attorney and the Patent Office. File wrappers are open to public inspection and are very revealing documents. They include all the amendments that had to be made to the original claims before any were finally allowed. The limitations on the issued claims will thus become apparent. It is often surprising to discover how exceedingly narrow the claims of an otherwise impressive-appearing patent may actually be. As a general rule, the longer a claim is, the more restricted it is likely to be.

Invention agreements

It has become an almost universal practice to require technical personnel to sign an invention-assignment agreement as one of the con-

---

[19] Berle and de Camp, *op. cit.*, p. 61.

ditions of employment. Such agreements may vary in length, but all generally amount to the following:

1. The employee agrees to disclose to his employer all inventions he may make that fall in his employer's line of business and to assign these inventions to his employer during his term of employment.
2. The employee agrees to perform all acts necessary in applying for patents and in executing assignments.
3. The employee agrees to maintain secrecy on all knowledge gained during the course of his employment.[20]

With regard to point 1, note that it makes no difference whether the invention is made on the employer's time or on the employee's. Technical personnel are *expected* to make inventions; it is one of the things they are hired for. It would be ridiculous to permit a man to withhold an invention from his employer just because his idea came in the middle of the night. If this were possible, who could ever prove that *all* ideas did not occur in the middle of the night?

For nontechnical personnel, the situation is usually a little different. If the employer's time or facilities were used in making the invention, then the employer would have a royalty-free "shop-right" to use it, with title remaining in the employee's hands. If everything were done on the employee's time and without the use of any equipment belonging to the employer, then generally, all rights would remain with the employee.

Returning to the case of the technical employee, the next question is, "What happens if the invention is *not* in the company's line of business?"

If the invention is remote from the company's present line and if all work was done without the use of anything belonging to the employer, then any resulting patent will belong to the employee. However, if the employee attempts to sell his invention, he may discover that the prospective purchaser will demand a release from the inventor's employer, showing that he has a right to offer the invention for sale. This could be a sore point because the company has no obligation to grant releases unless the employment agreement says so. As a further point, if the employee has made any inventions *prior* to employment and he wishes to retain clear title to them, he should make sure that they are listed, described, and specifically exempted from his employment agreement.

Sometimes an employment agreement will include a clause that extends the terms of the agreement six months or a year beyond termina-

---

[20] F. H. Crews, "That Patent Agreement—Should You Sign? *Patent Problems*, 1961, pp. 23–25. (Booklet publ. by *Product Engineering* magazine, McGraw-Hill, New York.)

tion of a man's employment. Many attorneys consider such clauses to be of doubtful legality, and proposals have been made in Congress that contracts of this sort be prohibited. Even so, prior to passage of a law rendering such clauses illegal, their attempted enforcement could result in considerable unpleasantness, cost, and inconvenience.

An agreement to maintain secrecy is of a different nature. In fact, an employee is legally bound to maintain secrecy about his employer's trade secrets, both during and after his term of employment, even though he has not signed any explicit agreement to that effect.[21]

Confidential disclosure

There are hidden legal hazards in accepting ideas from outsiders. If there is nothing to indicate otherwise, the law has often taken the position that such disclosures are given in confidence and that, therefore, the ideas may not be used without permission and payment.[22] If a corporation does not take appropriate protective steps, it could even find itself in the irritating position of having to pay for an idea it had already developed for itself, if it accepts suggestions from outsiders. Worse yet, it might be made to pay for something that is already well known and in the public domain if it permits a relationship to develop which makes it appear that the submitter was entitled to payment.

Large corporations sometimes almost feel besieged by outside suggestions. Once in a great while, a really useful suggestion may come along, but it is astounding how spurious and trivial most of the ideas are. During the NAS symposium previously referred to (see footnotes 11 and 15, this chapter), a company executive of the Polaroid Corporation stated that only 3 out of 2000 unsolicited ideas that had been submitted to his corporation proved to have any value and that even those 3 did not amount to much.[23]

Unsolicited (and presumably confidential) disclosures may be made in a myriad of ways. They may casually come up in social conversations and can create just as much legal trouble as those that arrive through more formal routes. Engineers, especially, must be alert to this possibility, since presumably, an engineer is in a qualified capacity to recognize and to make use of a good idea. The best legal advice, of course, is to prevent the disclosure from being given, if this is possible.

---

[21] J. H. Munster, Jr., and J. C. Smith, "The Care and Feeding of Intellectual Property," *Science*, May 7, 1965, pp. 739–743.

[22] "Corporate Protective Devices in the Acquisition of Ideas," *Harvard Law Review*, 1952, pp. 673–685.

[23] *The Role of Patents in Research—Part II, op. cit.*, p. 185.

Most commonly, unsolicited disclosures arrive in the mail and the corporation is unaware that it has a confidential disclosure on its hands, until the envelope has been opened and its contents examined. A widely used protective procedure is to stop reading and to re-enclose the material in its envelope as soon as its nature is recognized, and then send it to the company's legal department. The presumption is that the legal department is too far removed from the operational end of the business to be able to apply the idea. The material is retained by the legal department (so that the company will later be able to prove exactly what was submitted), and a form letter is sent to the submitter explaining, in a friendly fashion, that the company *is* interested in receiving ideas, but only if they are accompanied by a signed release. A release form is included, which usually stipulates, among other things, that no relation of confidence exists and that the parties agree to abide by only such rights as may properly arise from issued patents.[24]

It may appear that the submitter is being asked to throw himself upon the mercy of the corporation, and this is approximately the situation. However, the corporation is actually the more vulnerable of the two parties, and so must protect itself.[25]

---

[24] See C. G. Baumes, *Patent Counsel in Industry* (New York: National Industrial Conference Board, 1964) for more detail.

[25] For some details of actual cases, see C. D. Tuska, *Inventors and Inventions* (New York: McGraw-Hill, 1957), pp. 159–168; and R. A. Buckles, *Ideas, Inventions, and Patents, op. cit.*

FOURTEEN

Engineering societies

Nearly 200 engineering societies or related groups are listed in the *Directory* of the Engineers Joint Council, not counting state and local organizations. In addition, there are another 15 engineering societies in Canada, and approximately 21 international bodies. Probably, no other profession is organized into such a number of societies.[1]

The remarkable proliferation of engineering societies often leads engineers to ask if some degree of unity is not desirable and possible. Actually, several major bodies work toward attaining unified action in the engineering profession; prominent among these associations is the Engineers Joint Council, which carries out numerous programs for the benefit of 485,000 members of its affiliated societies. Moreover, the National Academy of Engineering, the Engineers' Council for Professional Development, the United Engineering Trustees, and the National Society of Professional Engineers are all concerned with the engineering profession as a whole.

Purpose

Most American engineering societies exist for the basic purpose of disseminating information. As an example, some purposes of the American Society of Civil Engineers (ASCE), this country's oldest engineering society, are given here:

1. To encourage and publicize discoveries and new techniques throughout the profession
2. To afford professional associations and develop professional consciousness among civil engineering students

---

[1] *Directory of Engineering Societies and Related Organizations* (New York: Engineers Joint Council, January 1970).

3. To further research, design and construction procedures in specialized fields of civil engineering
4. To give special attention to the professional and economic aspects of the practice of engineering
5. To enhance the standing of engineers
6. To maintain and improve standards of engineering education
7. To bring engineers together for the exchange of information and ideas

Most of the principal engineering societies have objectives similar to those of the ASCE. To fulfill the purpose of disseminating and publicizing information, the societies publish technical journals and hold numerous technical meetings at the national, regional, and local level. For example, during 1972, the ASCE sponsored 11 national meetings, with 270 technical and professional sessions. The American Society of Mechanical Engineers (ASME) was even more active, sponsoring 29 national meetings during 1972, with more than 1000 technical papers. National meetings of the major societies regularly attract 3000 or 4000 engineers. The 1972 Offshore Technology Conference, sponsored jointly by several societies, drew 22,000 registrants. In 1971, our largest society, The Institute of Electrical and Electronics Engineers (IEEE), was publishing no less than 31 different technical journals (referred to collectively, as the *IEEE Transactions*). Some major engineering societies actually bear a strong resemblance to big publishing houses and maintain large full-time staffs principally to handle these publishing activities, at the United Engineering Center in New York. As an example, the IEEE published a total of 196 different issues of its *Transactions* and other journals in 1971.

Numerous personal advantages are derived from membership in a professional society, especially membership in one of the five major societies referred to as the *Founder Societies*. These Founder Societies are:

American Society of Civil Engineers
American Institute of Mining, Metallurgical, and Petroleum Engineers
American Society of Mechanical Engineers
The Institute of Electrical and Electronics Engineers
American Institute of Chemical Engineers

The Founder Societies have been given this name because in 1904 they founded United Engineering Trustees, Inc. (of which more will be said later). They all require that a man meet high standards before he can become a member, and they account for a combined membership of 378,000 engineers (including some duplicate memberships).

Some personal advantages of membership relate to the regular receipt of journals and periodicals and to attendance at society meetings at a reduced cost. More important, it has been observed that members of professional societies have higher average earnings than nonmembers, although this particular statistic can probably be attributed to the fact that, the more natural ability an engineer has, the more likely he is to join a society. The most important reason for joining one or more professional societies, however, is simply this: if the various societies had never existed, the progress of the engineering profession (and, therefore, of all civilization) would be far behind its present position; the societies cannot exist or function unless engineers belong to them and support them; it follows, then, that the fundamental reason for belonging to a society is *to participate* in the activities of the profession and in the producing of the social benefits that flow from these activities.

Membership

Membership grades differ somewhat among the societies, but not significantly so. The ASME grades of membership are reasonably typical and, therefore, are briefly described in the following paragraphs:

STUDENT MEMBER. Belongs to a student chapter at a school with a curriculum approved by the Engineers' Council for Professional Development (ECPD).

ASSOCIATE MEMBER. Must have graduated from an engineering school of recognized standing, or have eight years of acceptable experience. No entrance fee for graduate members of student chapters.

MEMBER. Requires six years of practice if a man was graduated from an engineering school of recognized standing; otherwise, 12 years are required. He must have spent five years in a capacity where he was in responsible charge of work.

FELLOW. Requires nomination by fellow members for an engineer or engineering teacher of acknowledged attainments. Also requires 25 years of practice, and 13 years in Member grade. This is an honorary grade.

HONORARY MEMBER. Requires nomination by membership and election by the Board of Direction. This grade is given to persons of distinctive engineering accomplishment.

EXECUTIVE AFFILIATE. Not necessarily an engineer, but in a position of policy-making authority relating to engineering; he must be a man who cooperates closely with engineers.

AFFILIATE. Capable and interested in rendering service to the field of engineering.

Engineers Joint Council (EJC)

Seventeen American engineering societies are members of the EJC (see Table 14-1). This body acts as the representative of the engineering profession as a whole, in cases where such representation is appropriate, and helps develop public policy regarding the profession.

One of EJC's most important components is the Engineering Manpower Commission (EMC), a group that conducts many studies, including investigations on efficient engineering-manpower utilization; the future demand for engineers, scientists, and technicians; and the placement of engineering graduates. The EMC regularly conducts the most useful of all engineering salary surveys. The EJC has also aided in the interpretation of engineering through the public media and has provided personnel to testify before governmental, professional and industrial groups.

Engineers' Council for Professional Development (ECPD)

The ECPD is made up of 14 participating organizations as follows:

American Institute of Aeronautics and Astronautics (AIAA)
American Institute of Chemical Engineers (AIChE)
American Institute of Industrial Engineers (AIIE)
American Institute of Mining, Metallurgical, and Petroleum Engineers (AIME)
American Nuclear Society (ANS)
American Society of Agricultural Engineers (ASAE)
American Society of Civil Engineers (ASCE)
American Society for Engineering Education (ASEE)
American Society of Mechanical Engineers (ASME)
The Institute of Electrical and Electronics Engineers (IEEE)
National Council of Engineering Examiners (NCEE)
National Institute of Ceramic Engineers (NICE)
National Society of Professional Engineers (NSPE)
Society of Automotive Engineers (SAE)

The ECPD concerns itself principally with 1) the accreditation of engineering curriculums, 2) the professional development of young engineers, and 3) the formulation of the Canons of Ethics.

The most visible activity of the ECPD is its work in curriculum accreditation. In order to become accredited, an engineering school must

Table 14–1 Societies affiliated with the Engineers Joint Council or the Engineers' Council for Professional Development[a]

Name of society	*Location*	*Members*	*Year founded*
Member Societies of EJC			
American Society of Civil Engineers (ASCE)	UEC,[b] New York, N.Y. 10017	67,427	1852
American Institute of Mining, Metallurgical, and Petroleum Engineers (AIME)	UEC, New York, N.Y. 10017	45,800	1871
Constituent Societies of AIME:			
Society of Mining Engineers			
Society of Petroleum Engineers			
The Metallurgical Society			
American Society of Mechanical Engineers (ASME)	UEC, New York, N.Y. 10017	60,682	1880
American Society for Engineering Education (ASEE)	Washington, D.C. 20036	12,350	1893
Society of Naval Architects and Marine Engineers (SNAME)	New York, N.Y. 10006	9,038	1893
American Society for Testing and Materials (ASTM)	Philadelphia, Pa. 19103	16,500	1898
American Society of Agricultural Engineers (ASAE)	St. Joseph, Mich. 49085	7,244	1907
American Institute of Chemical Engineers (AIChE)	UEC, New York, N.Y. 10017	39,252	1908
American Institute of Consulting Engineers (AICE)	UEC, New York, N.Y. 10017	420	1910

Table 14–1 (continued)

Name of society	Location	Members	Year founded
American Society for Metals (ASM)	Metals Park, Ohio 44073	39,855	1913
Society of American Military Engineers (SAME)	Washington, D.C. 20006	26,000	1919
Society for Experimental Stress Analysis (SESA)	Westport, Conn. 06880	2,669	1943
Instrument Society of America (ISA)	Pittsburgh, Pa. 15219	20,400	1945
American Institute of Industrial Engineers (AIIE)	UEC, New York, N.Y. 10017	17,540	1948
Society of Fire Protection Engineers (SFPE)	Boston, Mass. 02110	1,400	1950
American Institute of Plant Engineers (AIPE)	Cincinnati, Ohio 45208	4,026	1954
American Association of Cost Engineers (AACE)	Morgantown, W. Va. 26506	1,880	1956
Members of ECPD but not of EJC [c, d]			
The Institute of Electrical and Electronics Engineers (IEEE)	UEC, New York, N.Y. 10017	163,940	1884
Society of Automotive Engineers (SAE)	New York, N.Y. 10001	29,362	1905
National Council of Engineering Examiners (NCEE)	Clemson, S.C. 29631	—	1920
American Institute of Aeronautics and Astronautics (AIAA)	New York, N.Y. 10019	23,072	1931
National Society of Professional Engineers (NSPE)	Washington, D.C. 20006	69,500	1934
National Institute of Ceramic Engineers (NICE)	Columbus, Ohio 43214	1,651	1938
American Nuclear Society (ANS)	Hinsdale, Ill. 60521	10,000	1954

[a] Information given in this table has been taken from the *Directory of Engineering Societies and Related Organizations, 1970* (New York: Engineers Joint Council, 1970).

[b] UEC stands for United Engineering Center.

[c] The list of members of the ECPD is taken from the *41st Annual Report: 1972–73*. (New York: Engineers' Council for Professional Development, Sept. 30, 1973).

[d] Additional members of the ECPD are the AIChE, AIIE, AIME, ASAE, ASCE, ASEE, and ASME.

Table 14–2 Other large societies

Name of society	*Location*	*Members*	*Year founded*
American Water Works Association (AWWA)	New York, N.Y. 10016	21,500	1881
American Society of Heating, Refrigerating, and Air-Conditioning Engineers (ASHRAE)	New York, N.Y. 10017	24,250	1894
American Welding Society (AWS)	New York, N.Y. 10017	20,895	1919
Society of Manufacturing Engineers (SME)	Dearborn, Mich. 48128	42,452	1932
American Society for Quality Control (ASQC)	Milwaukee, Wisc. 53203	23,296	1946
Association for Computing Machinery (ACM)	New York, N.Y. 10036	25,747	1947

Source: *Directory of Engineering Societies and Related Organizations, 1970* (New York: Engineers Joint Council, 1970).

Table 14–3 Engineering honoraries

Name of society	*Location*	*Members*	*Year founded*	*Remarks*
Tau Beta Pi	Univ. of Tennessee, Knoxville, Tenn. 37916	152,000	1885	national engineering honorary
Society of the Sigma Xi	New Haven, Conn. 06510	104,000	1886	recognition of noteworthy achievement in research
Eta Kappa Nu	Okla. State Univ., Stillwater 74074	60,000	1904	national electrical engineering honorary
Pi Tau Sigma	Univ. of Ill., Urbana, Ill. 61803	43,000	1915	national mechanical engineering honorary
Chi Epsilon	Austin, Tex. 78712	24,543	1922	national civil engineering honorary
Alpha Pi Mu	Virginia Poly. Inst., Blacksburg, Va. 24060	7,000	1949	national industrial engineering honorary

Source: *Directory of Engineering Societies and Related Organizations, 1970* (New York: Engineers Joint Council, 1970).

have already produced a number of graduates and must undergo a careful inspection of its program by an accreditation review team, which is usually composed of engineering educators. Accreditation is granted by individual *curriculum*, and not for a school or a college as a whole. In 1973, 228 American institutions had one or more engineering curriculums approved by the ECPD.

Curriculums in engineering technology also are accredited by the ECPD, and in 1973, 92 American institutions had such accredited curriculums. Most of these programs were at the two-year level, but 24 institutions had accredited engineering technology curriculums at the four-year level.

National Council of Engineering Examiners (NCEE)

The NCEE's membership consists of all members of the state boards of registration of the 50 states, the District of Columbia, the Canal Zone, Guam, Puerto Rico, and the Virgin Islands—55 boards in all. The NCEE works toward improving engineering registration laws, especially with regard to establishing uniformity among the states and interboard recognition of engineer registration. The Council meets once each year and publishes a quarterly *Registration Bulletin*, in addition to the *Proceedings* of its annual meeting.

National Society of Professional Engineers (NSPE)

The NSPE has, as its primary objective, ". . . the promotion of the profession of engineering as a social and an economic influence vital to the affairs of men and of the United States." As one means to this end, it has its headquarters in the national capital and takes a direct interest in legislation that affects the engineering profession. The NSPE was founded in 1934 and, as of 1973, was America's second largest engineering society, with approximately 69,000 members. The annual *Engineers' Week* constitutes one of the activities of the NSPE.

Fifty-three different state societies (such as the Illinois Society of Professional Engineers and the California Society of Professional Engineers) are member state societies of NSPE. Membership in a state society automatically makes one a member of the NSPE.

United Engineering Trustees, Inc.

The engineering profession owns a magnificent 20-story *United Engineering Center* on United Nations Plaza, in Manhattan. The center was built with the help of $5 million donated by industry and $4 million by individual members of the profession and today is debt-free.

It was completed in 1961 and provides a home for 23 engineering societies and related organizations. The United Engineering Trustees, acting for the profession, is the titular owner of the *United Engineering Center.*

Other activities of the United Engineering Trustees are the

ENGINEERING SOCIETIES LIBRARY—a free public engineering library containing 220,000 volumes relating to all branches of the profession.

ENGINEERING FOUNDATION—an endowed foundation for "the furtherance of research in science and engineering, and the advancement in any other manner of the profession of engineering and the good of mankind." The foundation has assisted in the stimulation of many research programs over the years, and in the formation of the National Research Council (National Academy of Sciences), and the National Academy of Engineering.

National Academy of Engineering (NAE)

It could almost be said that the engineering profession became of age on December 11, 1964, with the announcement that a National Academy of Engineering had been formed. As *Saturday Review* commented: ". . . the half million or more engineers in the country had achieved a voice in public affairs on a prestigious level equal to that of the country's quarter million scientists."[2]

The National Academy of Engineering (NAE) was set up by the joint efforts of the EJC, the Engineering Foundation, and the National Academy of Sciences (NAS), under the original congressional charter granted to the National Academy of Sciences in 1863. The two academies are autonomous and parallel, but spokesmen from both groups have declared their intention to operate them on a closely coordinated and cooperative basis.

Election to the NAS is one of the highest honors an American scientist can receive, and the same is true of the NAE for engineers. In 1970, the total membership of the NAS was 843, and of the NAE, 279.

The purpose of the Academy is not only to honor outstanding engineers, but also to provide a body of expert engineering knowledge that can contribute strong guidance for the nation's technological affairs. The purposes of the NAE briefly are as follows:

1. To provide means for assessing the constantly changing needs of the nation and the technical resources that can and should be applied to them

---

[2] *Saturday Review*, February 6, 1965, p. 51.

2. To explore means for promoting cooperation in engineering in the United States and abroad
3. To advise Congress and the executive branch of the government on matters of national import pertinent to engineering
4. To recognize outstanding contributions to the nation by leading engineers

Mergers between societies

Engineers have been remarkably active in creating new societies, but there have also been some notable mergers in recent years: In 1963, the American Institute of Electrical Engineers (AIEE) combined its membership of 58,000 engineers with that of the Institute of Radio Engineers or IRE (about 93,000), to create the Institute of Electrical and Electronics Engineers (IEEE). The AIEE had been founded in 1884, and the IRE in 1912. Also in 1963, the Institute of Aerospace Sciences (IAS) and the American Rocket Society (ARS) merged to form the American Institute of Aeronautics and Astronautics (AIAA). The ARS was founded in 1930, and the IAS in 1932.

FIFTEEN

Professional registration

Most people are surprised to discover that the first state to adopt a registration law (in 1907) was not a big industrial state, but Wyoming. That state's registration law was passed to protect the public from a flood of persons of doubtful qualifications, who were representing themselves as engineers during an era of great water-resources development.

Today, all 50 states, the District of Columbia, and four territories have engineering registration laws. Unfortunately, there are differences in the requirements for registration in the various jurisdictions. Lack of uniform qualifications has been a major barrier to the effectiveness of registration. Proposals have occasionally been put forth for national registration, but without success because regulation of the professions is a constitutional right reserved to the states.[1] Nevertheless, the registration movement has grown steadily and reached a total of 393,725 registered engineers in 1972. (It has been estimated that this total represents only about 250,000 individual licensees, because of the widespread practice of multiple registration in several states.)[2]

The public case for registration

Protection of the public was the justification for the passage of our first registration law and is, today, still the only justification for having such laws, from the viewpoint of public policy. The case for registration has been most eloquently stated by a Utah court of law, as follows:

> It has been recognized since time immemorial that there are some professions and occupations which require special skill, learning, and

---

[1] O. B. Curtis, Sr., "Reciprocity: New Look at Old Subject," *NCEE Registration Bulletin,* July 1973.

[2] *1972 Proceedings* (Clemson, S.C.: National Council of Engineering Examiners, 1973), pp. 57, 210.

> experience with respect to which the public ordinarily does not have sufficient knowledge to determine the qualifications of the practitioner. The layman should be able to request such services with some degree of assurance that those holding themselves out to perform them are qualified to do so. For the purpose of protecting the health, safety, and welfare of its citizens, it is within the police power of the State to establish reasonable standards to be complied with as a prerequisite to engaging in such pursuits.[3]

In a 1963 decision, a Delaware court quoted the preceding passage, in applying it to engineering, saying, "Professional engineering is recognized as one such occupation."[4]

It is commonly taken for granted that state engineering-registration laws apply only to those engineers who offer their services directly to the public. In many states this is indeed the case, and the laws in those states contain clauses that specifically exempt engineer employees of manufacturing companies from registration. But some states do not have such exemption clauses in their laws; and engineers in those states, who assume they are exempt might find themselves in an equivocal position.[5] The "Model Law," prepared by the National Council of Engineering Examiners (NCEE) as a guide for state law-making bodies does not contain such an exemption. In fact, in 1972 the NCEE voted to press for changes in registration laws which would specifically include engineers engaged in the design of ". . . industrial or consumer products or equipment of a mechanical, electrical, hydraulic, pneumatic, or thermal nature, insofar as they involve safeguarding life, health, or property."[6]

The ambiguity of the current situation is best demonstrated by a consideration of the scope of protection afforded the public by registration. It has been adroitly pointed out by A. W. Weber, that registration is almost universally required for the engineering of *static* items such as bridges, highways, and buildings, but almost never for *dynamic* items such as automobiles, locomotives, and airplanes. Weber points out that there is far more danger of death and destruction associated with the latter group, yet such items are designed substantially by non-registered engineers.[7]

Engineers often look at the medical profession with envy, wishing

---

[3] *Clayton v. Bennett*, 298 P. 2d 531.

[4] *State of Delaware v. Frank N. Durham*, 191 A. 2d 646 (1963).

[5] M. F. Lunch, "Engineering Registration: The Legal Opinion," *Mechanical Engineering*, May 1964, p. 23ff.

[6] M. F. Lunch, "Product Design: Engineering Registration Laws," *Professional Engineer*, November 1972, p. 18ff.

[7] A. W. Weber, "Licensing: The Engineer's Dilemma," *Mechanical Engineering*, May 1965, p. 116.

their own profession possessed the same degree of public identity and esteem. Some have concluded that universal registration is the key to the doctors' prestige and they seek a similar course for engineers. It should be observed, however, that a fine line exists between legislation that protects the public and legislation that *protects the profession.* It is a nice exercise in semantics to distinguish between 1) moves that enhance public esteem for the engineering profession in order to increase the confidence in which the public may seek the service of engineers and 2) moves that enhance public esteem for the engineering profession in order to increase the financial position and personal prestige of engineers. The former motive has a proper place in public policy; the latter does not.

Some groups within the profession have taken the viewpoint that an unregistered engineer is simply not a member of the profession. However, it has not yet been possible to equate registration with professionalism. Far too many eminent engineers are not registered, and far too many of those engineers who *are* registered, achieved that status by routes which required neither a written examination nor a college degree.

A major issue in engineering registration concerns license renewal. Some state boards of registration have considered whether an engineer should be required to certify that he has undertaken a certain minimum amount of continuing education or other professional development in order to have his license renewed. A similar issue faces all occupations that have licensing boards, including the medical and law professions. Many engineers have opposed a law of the foregoing type on the grounds that no demonstration of need has been shown, or that engineers are failing to meet their responsibilities under present licensing laws. The issue was unresolved at the time of writing this book, but it appeared likely that laws of the type under consideration would start making their appearance during the 1970s.

The personal case for registration

In spite of the problems and inconsistencies in registration, there are compelling reasons why every young engineer should become registered as quickly as possible after graduation. Among these reasons are the following:

1. Registration may become more important in the future.
2. No one can foresee the future course of his own career: an individual may believe his career will lie exclusively in areas not requiring registration; but he could change his mind, or unexpected opportunities may arise for which registration is required, or activities currently unaffected by the law might become affected in the future.

3. A court of law generally will not recognize an individual as an engineer, unless he is registered; this could be frustrating, should the engineer want to testify as an expert witness for example.

4. If a nonregistered person engages in practice required by law to be performed by registered engineers, then at the very least, he may find that the courts will not aid him in attempts to collect his fee. "A statute requiring physicians, lawyers, or engineers to be licensed makes a contract by such persons, if not licensed, void."[8]

5. Many companies believe it is desirable for members of their engineering management to be registered; hence, registration could be an aid in promotion.

6. As the passage of a written 16-hour examination becomes more universal as a requirement, registration will undoubtedly be increasingly regarded by employers as an indicator of technical competence in a man seeking to be hired.

Corporate practice

A surprising way in which engineers employed by manufacturers could find themselves in conflict with state laws arises from what is called "corporate practice." Many of the states allow corporations to practice engineering, provided their principals or officers are registered. This is usually presumed to apply to instances wherein corporations offer engineering services to the public. But suppose Corporation "A," primarily a manufacturing company, offers to design and construct an item for Corporation "B," which will then be used by Corporation "B" in something it is constructing—in a power plant, for example. It is possible that Corporation "B" might be construed as a member of the "public," and then Corporation "A" would be violating the law unless its principal engineers were registered. Even the practice of engineering for internal purposes of a corporation might be in question. Many states make provision, in their laws, for unregistered persons (presumably intended to include corporations) to practice engineering on their own property, provided public safety is not involved. Other states definitely prohibit such practice, while the remainder make no provision one way or the other. In this connection, courts of law might have to decide the meaning of "public safety." If groups of visitors regularly are taken on tours through an industrial plant, is public safety involved? Are employees of a company to be considered members of the "public," and does their regular use of the corporation's premises involve public safety? State laws are continually undergoing amendment to answer questions like these.

---

[8] D. T. Canfield and J. H. Bowman, *Business, Legal and Ethical Phases of Engineering* (New York: McGraw-Hill, 1954), p. 199.

One group of engineers clearly unaffected by registration laws are those who are totally concerned with research. Table 5-3 shows, however, that only about 8 percent of all engineers are so engaged. By the definitions given earlier in this book, engineers engaged principally in research would be called "engineering scientists," and no one has yet attempted to write a registration law for scientists.

Interstate practice

Most states provide for temporary permits for registered engineers from other states to practice engineering in their territories. In addition, nearly all of them provide for what is generally termed "comity"[9] (literally, friendliness; consideration for others); this means that a state board may register, without examination, an engineer who is registered in another state, provided he attained registration in his own state on the basis of comparable qualifications.

A practice misunderstood by many people is that of registration with the Committee on National Engineering Certification (NEC). This is often assumed to be equivalent to registration for all states, or to registration with the federal government. Such interpretations are incorrect, since the federal government does not register anyone. The NEC is one of the activities of the NCEE and was organized to facilitate interstate practice. Upon request and payment of certain fees, the NEC will prepare a Council Record of the applicant's background. If certain eligibility standards are met, the NEC will issue a Council Certificate. With the Council Certificate and Council Record, comity registration can be obtained in a majority of the states without further examination.[10]

Requirements for registration

The pattern of requirements for professional registration prevailing in most of the states is as follows:

> Graduation from an ECPD-accredited school, plus four years engineering experience acceptable to the board, plus a 16-hour written examination
>
> *(or)*
>
> Eight years of engineering experience acceptable to the board, plus a 16-hour written examination.

---

[9] Another term frequently used—though incorrectly—is "reciprocity."

[10] *Regarding NCEE Services, Council Records, and Certification* (Clemson, S.C.: National Council of Engineering Examiners).

Some states have requirements higher than the preceding, while a few have requirements that are substantially lower.[11] The general pattern is for the 16-hour written examination to be divided into two equal parts; the first is generally known as the "Fundamentals Examination" (sometimes referred to as the Engineer-in-Training exam, or the EIT) and the second, as the "Professional Examination," or the "Principles and Practices" exam.

Nearly all the states have made provision for an EIT status and will allow persons to take the first eight-hour (EIT, or "Fundamentals") portion of the written examination immediately before or immediately after graduation. EIT status conveys no legal privileges and is offered primarily as a convenience to new graduates, so that they may take the examination in fundamentals at a time when the material is fresh in their minds. California has an unusual provision in its regulations, which grants four years of engineering experience credit to anyone who passes the EIT examination. Almost all of the states utilize a uniform national EIT examination, administered through the NCEE, and the great majority also use a uniform national examination for the "Professional" portion.[12]

Questions often arise concerning experience that will be acceptable to state boards. In general, a board will accept *creative* engineering experience requiring the application of mathematics and the engineering sciences to the planning and design of engineering works. Types of experience that are generally *un*acceptable are 1) teaching of non-engineering courses; 2) periods of summer employment taken prior to graduation; 3) assignments in the armed forces that are nonengineering in nature; 4) sales work (except those portions that may be of a creative engineering nature); and 5) construction or contracting, unless it involves actual engineering.[13] It may take an individual more time than the stated minimum number of years of experience before he can collect a sufficient amount of experience of the kind acceptable to a given board.

Many states make provision for cases of "eminence," whereby highly competent engineers of many years of experience can become registered without having to meet all the stipulated requirements. However, these practices are highly variable and usually are at the "discretion of the board"; hence, they will not be discussed in detail here.

---

[11] *The Practice of Engineering in the United States* (Clemson, S.C.: National Council of Engineering Examiners, 1973).

[12] *NCEE Registration Bulletin*, October 1973, p. 15.

[13] J. D. Constance, *How to Become a Professional Engineer* (New York: McGraw-Hill, 1958), pp. 98, 127.

SIXTEEN

Engineers' unions

Unionized engineers are a very small fraction of all engineers. Even at the peak of engineering union activity (about 1956–1958), less than 10 percent of the total engineering population was organized into unions.[1] By 1963, it was estimated that union representation had decreased to the point where it accounted for less than 5 percent of all engineers.[2] Yet, the existence of engineering unions has caused an impact upon the public's mind completely out of proportion to the unions' size. To much of the public, the image of an engineer going on strike is highly incompatible with the image of the engineer as a professional who administers to the public welfare.

There are many ramifications to engineering unionism, but the basic question in most engineers' minds appears to be the one of professionalism versus unionism. Since most engineers' unions have claimed that it is both possible and proper for an engineer to be simultaneously a professional and a union member, it behooves us to look briefly at the record.

The rise of engineering unionism

Although some engineering unions existed prior to World War II, most of them were organized between 1945 and 1947, in response to three potent forces:

1. Engineers shared the uncertainty of virtually all employees about their continued security of employment.

---

[1] R. E. Walton, *The Impact of the Professional Engineering Union* (Cambridge, Mass.: Div. of Research, Graduate School of Business Administration, Harvard University, 1961), pp. 388–395.

[2] *Professional Responsibility vs. Collective Bargaining* (Washington, D.C.: National Society of Professional Engineers, November 1963).

2. Many engineers believed their pay had seriously lagged in comparison with that of production workers. Many attributed this to the activities of the unions working on behalf of production workers.[3]

3. Many groups of engineers organized themselves to prevent being swallowed up by regular labor unions. Several National Labor Relations Board (NLRB) decisions under the Wagner Act showed that professionals might be forced, against their wishes, into bargaining units containing nonprofessionals.[4]

At first, the major engineering societies actively urged engineers to form their own bargaining groups because they believed the engineering profession would be irreparably damaged by inclusion in traditional labor organizations. Many managements also encouraged this formation, believing that the interests of both engineers and management would be enhanced by it.

In 1947, the passage of the Taft-Hartley Act ensured that professionals would not be required to join heterogeneous labor organizations (those that lump professionals and nonprofessionals together) against their wishes. This bill eliminated one of the most important motivating forces toward unionization for engineers. Since 1952, there have been few major instances of groups of engineers voting in favor of unionization, but there have been several cases where engineering unions have been voted out of existence.

In organizing their unions, engineers have usually proclaimed, at the outset, that they intend to behave as professionals and to be led only *by* professionals. As will be seen, the failure of some of these groups to adhere to their initial declarations has been a factor leading to their ultimate rejection by the engineers they represented.

A 1959 survey among unorganized engineers and scientists in 10 well-managed companies showed that engineers' principal fears concerning unionism are 1) that salaries and promotions would not reflect individual contributions; 2) that professional productivity would be reduced; 3) that relations between professionals and management would be impaired; 4) that the objectionable policies and practices of labor unions might be introduced. Fifty percent of the engineers surveyed were *strongly* against engineers' unions, while only 5 percent were strongly in favor. However, many of those men opposing unions stated that their attitudes were based on the belief a union was "not necessary"; the implication was that their attitudes might change under different economic conditions.[5]

---

[3] Walton, *op. cit.*, pp. 18–22.

[4] H. N. Rude, "White Collar Unions and the Law," *Personnel Journal*, February 1964, p. 88ff.

[5] J. W. Riegel, *Collective Bargaining as Viewed by Unorganized Engineers and Scientists* (Ann Arbor: University of Michigan, 1959), p. 43.

The decline of engineers' unions

Apparently, engineering unions declined partly because of the favorable economic climate engineers enjoyed during the 1960s, and partly because of engineers' disillusionment when they discovered they could not have unions and still retain their professional ideals. Apparently, many early joiners of unions held the belief that their union could succeed in remaining dignified and "professional" and would act mostly as a discussion unit bringing problems to the attention of management. Some engineers have stated they joined unions only so that they could have a say in the activities of the organization that represented them.[6]

However, the basic purpose of a union is to employ forceful means to get an employer to do something he would not do otherwise. In his study of engineering unions, Walton has divided the history of engineers' union activity into the following periods:

1945–1947. In this phase, management and engineers collaborated in establishing engineers' "associations," in order to exclude other bargaining agents.

1947–1949. This phase brought one-way cooperation with management by the unions. Unions were ignored by management.

1949–1953. Beginnings of conflict occurred. Unions attempted to put teeth into their efforts at collective bargaining. Walton calls this period one of "containment-aggression."

1953–1958. For most unions, the pattern remained containment-aggression while a few entered a period of accommodation and cooperation. In some others, however, the pattern became one of outright conflict, marked by many strikes.

1957–1963. This phase took place after the publication of Walton's book, but it might be called one of union *decline*. Four major decertifications[7] of engineers' unions occurred during these years.

Through the rest of the sixties and early seventies, the attitudes of engineers toward unions were mostly characterized by indifference. Even in 1971, during a period when there was great worry over employment of scientists and engineers, two NLRB elections involving professionals resulted in votes for "no union."[8] In 1973, a major organizational drive

---

[6] J. Seidman and G. G. Cain, "Unionized Engineers and Chemists: A Case Study of a Professional Union," *Journal of Business*, July 1964, p. 238ff.

[7] The NLRB supervises elections by groups of employees, concerning whether they wish to be represented by a union, and if so, which union. If the vote is "yes," the NLRB "certifies" that the union selected by the group is its official bargaining agent. If the vote is "no union," this outcome is referred to as a "decertification."

[8] *Professional Engineer in Industry Newsletter*, April 1973, p. 4.

by the AFL-CIO's Oil, Chemical and Atomic Workers International Union (OCAW) was met mostly with apathy.[9]

Once organized, many unionized engineers decided that if they wished to force their demands upon an employer, they had to be willing to employ unions' traditional ultimate weapon—the strike. A three-day strike occurred in 1950, one of the first strikes to be carried out by an engineering union. The strike was unsuccessful, but five other major strikes occurred in the 1950s, some of which were considered by their participants to have been successful. Some tactics in addition to strikes were proposed, but not carried out; these were the refusal to submit engineering reports and the refusal to perform duties directly affecting production.[10] Such actions could hardly be considered professional.

One of the worst instances involving striking engineers lasted only six days, but featured engineers' marching on the picket line, snake-dancing, and egg throwing. Seven engineers were arrested and charged with disorderly conduct, plus assault and battery of a police officer.[11] The worst fears of the engineers in Riegel's study—that engineers' unions might adopt some of the objectionable practices of labor unions—had been fully realized.

Some other tactics that have been employed by engineers' unions are: referring to engineers who refused to cooperate in strikes as "scabs" and giving them the "silent treatment"; 24-hour telephone campaigns threatening and harassing those who would not walk the picket-line; after a strike, withholding essential technical information from non-members and feeding it to union members instead.

In their efforts to operate from positions of strength, engineers' unions several times have attempted to form (or join) national labor organizations. In 1952, nine engineer unions formed the Engineers and Scientists of America (ESA). In 1956, this organization reached its peak, but then decline set in as internal conflict developed over inclusion of nonprofessionals.[12] In 1960, after several disaffiliations and decertifications, ESA went out of existence.[13]

During the year 1958–1959, some of the groups that had split off from the ESA tried to organize the Engineers and Scientists Guild (ESG). The ESG would have allowed technicians and other nonprofessionals to join unions consisting predominantly of professionals. Apparently this movement was not successful, but another attempt, made in 1963, resulted in the formation of the Council of Engineers and

---

[9] "A Union for Industrial Scientists?" *Science*, September 14, 1973, p. 1030.

[10] Walton, *op. cit.*, pp. 28–31.

[11] "The Engineering Union—Dinosaur in the Space Age," *American Engineer*, July 1961.

[12] "Engineer Union Fights for Life," *Fortune*, May 1960, p. 246ff.

[13] *The Engineer in Industry in the 1960's* (Washington, D.C.: National Society of Professional Engineers, 1961), pp. 61–70.

Scientists Organizations—West (CESO-W), which claimed to represent 20,000 engineers and scientists.[14]

Another prominent trend of engineers' unions has been in the direction of affiliation with regular labor unions, notably the International Union of Electrical Workers (IUE), the United Auto Workers (UAW), and the American Federation of Technical Engineers (AFTE), all of which are AFL-CIO affiliates. Reportedly, traditional blue-collar unions are searching hard for ways to attract white-collar workers, a trend that has been attributed to the decrease in factory employment of production people. One important motive causing engineering unions to seek affiliation with the AFL-CIO is that the labor unions, upon occasion, have refused to support engineers' strikes unless the engineering unions joined a regular labor organization.[15]

The first major engineering-union decertification came about partly from the attempts of the union to affiliate with the UAW, in 1957. In choosing between the UAW, the ESA, and "no union," nearly two thirds of the 1400 professionals involved voted for "no union."[16]

Two major decertifications of engineering unions took place in 1960. One involved nearly 7000 engineers, in which the union was rejected by a vote of 3 to 2.[17,18] In 1963, two more decertifications occurred.[19]

The period of the aerospace buildup—roughly 1964 through 1967—was hardly a fertile time for union activity because engineers were so much in demand. But it was widely assumed that the 1969–1971 period, marked as it was by aerospace layoffs and widespread discontent among engineers, would usher in a fruitful season of union recruitment. Yet, no strong union movement emerged, apparently because of the deep suspicions many professionals have about unionism.[20]

Conflict: professionalism versus unionism

There is little doubt that unionism and professionalism are incompatible. Professionalism holds that the interests of society and of the client (or employer) are paramount. Unions are collective bargaining agents that sometimes place the economic interests of the members ahead of those of the client or employer. Unions have often been effec-

---

[14] "Unions for Engineers are Showing New Vitality," *Engineering News-Record*, April 11, 1963, pp. 82–83.

[15] *The Engineer in Industry in the 1960's*, *op. cit.*, pp. 64, 72–81.

[16] *Ibid.*, p. 64.

[17] "Engineers Say No to Union," *Chemical and Engineering News*, May 30, 1960, p. 27.

[18] *The Engineer in Industry in the 1960's*, *op. cit.*, pp. 81–85.

[19] *Professional Responsibility vs. Collective Bargaining*, *op. cit.*, p. 4.

[20] "Unionization: Scientists, Engineers Mull over One Alternative," *Science*, May 12, 1972, pp. 618–621.

tive in correcting harmful labor practices in the past. Nevertheless, the requirements of professionalism and of unionism may, by their very natures, be in direct opposition. The Engineers Joint Council, the major voice of the engineering profession, declared in 1956, "To the engineer who feels that life provides an opportunity for constructive contribution to society, collective bargaining with its attendant potentiality for creating conflicting obligations is not acceptable."[21]

As was mentioned in Chapter 4, some traditional labor unions have tried hard to convince engineers that they are not professionals, but have common interests with factory workers. However, surveys among engineers show that engineers perceive their role as entirely different from that of factory workers. Engineers want to be treated as individuals, with rewards based upon their individual contributions. They tend to identify with management and have a strong sense of idealism and professional responsibility. The *intangible* rewards (creative challenge and sense of accomplishment) are as important to most of them as are the tangible rewards. In Riegel's words, the engineer ". . . does not view his work merely as a way to earn a livelihood. To him it is a distinctive mission to which he has dedicated himself."[22] All these things make engineers difficult prospects for unionism. It takes considerable ineptness on the part of a company management to drive engineers to the conclusion that their only salvation lies in a union.

Union organizers blame the engineering schools and the professional societies for the engineers' lack of taste for unions. "Engineers are poisoned at the source—in the engineering colleges," is a charge ascribed to an AFL-CIO staff man.[23] One union official blames engineers' wives: "You would be surprised how much trouble we had with wives who simply were against having their husbands in unions."[24]

Walton's study showed that engineers in unionized companies have mixed feelings about unions. He studied 11 of the 12 engineering unions in the country that consist primarily of professionals; his conclusions were that the engineers in those organizations thought unionism was an uncomfortable solution at best.[25]

Unions have had some favorable results, of course, particularly in companies that have tended to treat their engineers with indifference. Unions have caused many managements to formulate more consistent and objective personnel practices (although management people almost universally find it necessary to deny that unions have had any influence

---

[21] *An EJC Report: Raising Professional Standards and Improving Employment Conditions for Engineers* (New York: EJC, 1956), p. 9.

[22] *Collective Bargaining as Viewed by Unorganized Engineers and Scientists, op. cit.*, p. 89.

[23] *Fortune*, May 1960, p. 251.

[24] *American Engineer*, July 1961.

[25] Walton, *op. cit.*, p. 375.

whatsoever, according to Walton). Unions have also tended to protect their members from capricious actions by poor supervisors. It could be argued, however, that the managements in such cases brought the unions upon themselves, by suffering the existence of poor supervision and poor personnel practices in their companies in the first place.

In two important areas—salaries and job security—the unions have experienced mostly frustration. In some cases, engineering unions have found that the pay raises they obtained after much painful negotiation have been matched, or even exceeded, in nonunionized companies.[26] Since between 90 and 95 percent of the country's engineers are non-unionized, this can be a scant source of comfort to the unions. The point is that companies have found it necessary to retain their salaries at competitive levels, regardless of union activity. In some instances, companies have even granted increases on their own volition, which has been a source of embarrassment and irritation for the unions. In one case, a company voluntarily introduced an educational assistance program for its engineers. The union actually complained that the company's action constituted an unfair labor practice, since it had not been subject to the negotiation process.[27]

Some unions have made overtime a special issue and have been successful in obtaining time-and-a-half pay for overtime. This has tended to put many engineers in a difficult position: they appear to want to be treated both different from, and the same as, production people. Time-and-a-half is guaranteed to production workers by federal acts, but engineers are specifically exempted from such provisions because they are professionals. Such exemption has made it possible to treat engineers separately from other employees and to give them more freedom (freedom from time clocks, for example, and time off for personal business). Virtually everyone expects that an engineer will upon occasion have to work overtime, without extra pay, when the fulfillment of his responsibilities requires it; what the unions have aimed their fire at is the supervisors' subtle pressuring of the men under them to give "free overtime." Such pressure occurs, for example, when it becomes evident to the work group that merit increases and promotions go primarily to those who work plenty of free overtime. No one, including the unions, has as yet been able to draw a clear line between "proper" overtime, which is important and necessary, and "improper" overtime, which is a condition for advancement. Insofar as the latter practice has been curbed by union pressure, the results would undoubtedly be considered salutary, by most people.

Two principal fears that have been expressed concerning engineering unions are that 1) seniority would be made to take precedence over ability and 2) engineers would be pigeon-holed into narrow job classifi-

---

[26] Walton, *op. cit.*, p. 55.
[27] *Ibid.*, p. 90.

cations. Neither fear actually materialized to the extent anticipated, but it could be argued that this failure at least partly results from the favorable engineering employment climate prevalent in the United States in most years since 1950.

Generally, unions have sought to have seniority included as a factor in layoff situations, but only after ability has first been taken into account. The measurement of ability, however, is the perfect breeding-ground for contention. Walton claims that, in such circumstances, many supervisors have retreated from any attempts to measure ability by subjective standards and have tended to rest their judgment of "ability" on objective items such as degrees and years of experience. Furthermore, in a layoff, a company may find it needs certain special skills more than others and may wish this factor to outweigh both seniority and ability. The unions have attempted to regulate this by forcing companies to maintain periodically up-dated ability lists and special skill lists, with no clear-cut result except conflict and frustration.

The efforts just mentioned have tended to work in the direction of realizing a main fear of engineers concerning unionism: pigeon-holing in job classifications. A union cannot hope to enforce its demands concerning layoff priority, unless it can clearly establish the categories within which ability and seniority will operate. In some instances, such trends have led to jurisdictional disputes, including attempts to limit the amount of engineering work an engineering supervisor can engage in, and complaints by draftsmen and technicians that engineers are performing work which rightfully belongs to them.[28]

Some managements report that the division of loyalty between union and company has caused them to limit the amount of confidence they can place in their engineers. They state that they have been hesitant to ask unionized engineers to sit in on policy meetings or to inform such engineers of confidential matters, for fear that the knowledge the engineers gain will be used against management as a union weapon. They have also been reluctant to grant unionized engineers full authority in dealing with customers.[29]

Some engineers' unions have tended to limit their efforts to the stabilizing of personnel policies, the bringing of issues into the open, and the publication of accurate salary information. Thus, they have actually operated much like the "sounding board" approach, favored by the National Society of Professional Engineers, as an alternative to unionization. Sounding boards are intended to promote communication beween engineering employees and top management; they serve as objective fact-finding bodies and are strictly nonbargaining in nature.[30]

The possibility that engineers will some day be forced to choose be-

---

[28] Walton, *op. cit.*, pp. 118, 181–242, 287–289.

[29] *Ibid.*, p. 294.

[30] *The Engineer in Industry in the 1960's*, *op. cit.*, p. 96.

tween professionalism and unionism remains a cause for concern. It should be remembered that many engineers have stated, in surveys, that they were opposed to unions because they were "not necessary." Riegel concluded from his studies that engineers might feel themselves compelled to form unions any time they were subjected to great job frustration, impersonal treatment, or believed they had suffered in salary.[31]

---

[31] Riegel, *Collective Bargaining as Viewed by Unorganized Engineers and Scientists, op. cit.*, p. 41.

SEVENTEEN

Engineering education

Probably, more changes in engineering education have occurred from 1955 to 1965 than in any other decade since the late 1800s. In 1955, the much praised and sometimes maligned *Report on Evaluation of Engineering Education*[1] (often called by its familiar but unofficial title, the Grinter Report) was published; this report proclaimed it was time for a major change in engineering education. Engineering, said the report, must be made more scientific—and more scientific it has become. Then, in 1965 appeared the Preliminary Goals Report,[2] which reaffirmed the findings of the Grinter Report and called for the next big change: it was time for more official recognition of the role of graduate study, especially the master's degree program, in preparing people for entry into the profession.

The 1955 report stirred up a hornet's nest when it was published, and the 1965 report did the same. Critics of the 1955 report declared that too much attention to science would deprive engineering of that which had made the profession great—its emphasis upon creative design—and would turn American engineering schools into second-class science departments. Critics of the 1965 Preliminary Goals Report worried that 1) a stampede to graduate school might be caused, perhaps resulting in a downgrading of the master's degree; or that 2) professional recognition might be denied to those who could not qualify for graduate school; or that 3) undesirable conformity might be imposed upon engineering education.[3]

In view of all the commotion, one would hardly want to claim that

---

[1] *Report on Evaluation of Engineering Education* (Washington, D.C.: American Society for Engineering Education, 1955). L. E. Grinter was chairman of the committee that prepared the report.

[2] *Goals of Engineering Education—The Preliminary Report* (Washington, D.C.: American Society for Engineering Education, 1965).

[3] "Goals of Engineering Education—Dissenting Opinions," *Mechanical Engineering*, January 1966, p. 40ff.

the engineering education picture is a static one. In fact, some find it so volatile that they almost despair of a solution. But, in the midst of the confusion, certain trends do appear:

First, engineering undeniably has become much more firmly based upon science and mathematics, and almost everyone would agree this is desirable.

Second, engineering education during the 1960s had *already* moved substantially into the graduate school, not because anyone decreed it should, but because individual engineers found it was necessary. By 1964, one third of all American graduates earning B.S. degrees in engineering were going on to master's degrees. By 1969, it was estimated that 56 percent of U.S. engineers below age 45 had advanced degrees of one form or other, and 65 percent of those between 25 and 35 had advanced degrees.[4]

Third, it would appear that engineering education had arrived at a reasonably satisfactory balance between theory and application, by the end of the 1960s. Most engineers were well enough equipped with a background of science and math so that they could meet the rapidly changing needs of modern society; yet, they seemed to understand once they moved into industry, that it is the *application* of science and math that counts for an engineer, and made the transition from college to industry without too much difficulty.

Because of the enormous protests against the recommendations of the Preliminary Goals Report, the final report appeared in 1968 with its recommendations couched in less controversial terms than in the Preliminary Report. Besides a general confirmation of the content of the Grinter Report, some of the principal recommendations of the final version of the Goals Report were:[5]

1. By the end of the 1970s, basic engineering education should be extended to include at least one year of graduate study leading to the master's degree. Graduate study should be recognized as an integral part of engineering education.

---

[4] A. A. Schultz, Jr., "Rationale for Accreditation of Advanced Professional Programs," presented at 39th Annual Meeting of ECPD, San Francisco, Calif., October 4, 1971. The figures are based upon a 1969 study conducted by the EJC which sampled the 345,000 memberships of 13 major engineering societies.

[5] *Final Report: Goals of Engineering Education* (Washington, D.C.: American Society for Engineering Education, January 1968). Chairman: E. A. Walker, Pennsylvania State University; Director, Graduate Phase of Study: J. M. Pettit, Stanford University; Director, Undergraduate Phase of Study: G. A. Hawkins, Purdue Universtiy.

2. Existing programs should be made more flexible. Many were perceived as too rigid to meet the needs of the future. Diversity should be encouraged.

3. Credit hour requirements should be reduced. It had become too difficult for engineering students to obtain a bachelor's degree in eight semesters.

4. The research function in engineering schools should be integrated with the educational purposes of the institution to the fullest possible extent.

5. Engineering colleges should provide high quality part-time advanced degree programs for engineers in industry and government, involving both on- and off-campus study.

6. ECPD should continue to focus its accreditation activities on the basic or "first professional" degree.

7. Continuing education programs should be provided to help in the maintaining of competence for professional engineers. These programs should supplement those part-time programs which lead to advanced degrees.

8. The engineering student should have sufficient acquaintance with the humanities and social sciences that can help him to understand the large social problems of his time. He should gain an appreciation of the importance of his own role in the solution of these problems.

The trend to graduate study

Historically, the length of education for the practicing professional engineer has been four years and has been carried out as an undergraduate operation. This is in sharp distinction to education in most of the other professions, which begins *after* a bachelor's degree has been granted.

In the past, graduate education in engineering has carried essentially the same implications as graduate education in science, that is, it was generally assumed that the man was preparing for a career in research or teaching. Full preparation for research or teaching meant going all the way to the Ph.D., but the master's degree was considered to be a step in that direction, since normal *professional* preparation stopped at the bachelor's degree.

Since World War II, however, more and more engineers have been going on to graduate degrees, and the Goals Report encouraged this trend. Nevertheless, the Goals Report also advised that universities continue to offer strong four-year bachelor's degree programs in engineering. Such programs would serve as excellent educational backgrounds for fields like management, sales engineering, operations, and contracting. The report pointed out that large numbers of men with the engineering B.S. degree enter these fields every year and that this trend is expected to continue. In fact, it has been asserted by some writers that engineering is actually more of a liberal education than the typical "liberal arts" curriculum, since engineering includes a large measure of humanities, as well as a great deal of science, whereas liberal arts curriculums seldom include much science and practically never include engineering.

Insofar as curricular content is concerned, the 1965 report reaffirmed the recommendations of the 1955 report and stated that engineering education should continue to be based strongly upon the physical sciences, the engineering sciences, and mathematics; the liberal education content should be strengthened and improved, and diversity and flexibility should be encouraged; *analysis, synthesis, and design* of systems should be given increased emphasis in engineering curriculums at all levels. The report stated that engineering educators ". . . must develop greatly improved programs which will stress creative design and development, give the student an appreciation of the importance of costs and an opportunity for experiencing the thrills of invention—the excitement of original and imaginative thought in his chosen field."

The report strongly urged *laboratory* experiences, because of the "feeling" for the actual physical situation laboratories can provide and because, by permitting evaluation of the performance of designs, such experiences may lead to the discovery of results not anticipated by theory.

In recommending the five-year master's degree program, the ASEE Preliminary Goals Report noted that several schools have offered five-year programs leading to the *bachelor's* degree in the past, but that such curriculums have, for the most part, disappeared. For example, in 1965, Cornell and Minnesota, two major engineering schools that had been prominent in the five-year bachelor's field, converted their five-year programs to yield master's degrees, instead.

Another special kind of graduate engineering program used by some schools leads to the degree of "Engineer." In past years, this degree has occasionally been awarded to practicing engineers who have completed five years of experience plus a thesis. According to the Goals Report, the foregoing mode of handling the Engineer degree has essentially disappeared and has been replaced by another, which treats the Engineer degree in the same fashion as the master's and doctor's degrees but places it in an intermediate position. In 1972, about 350 Engineer degrees were awarded across the nation, more than 60 percent of them by the Massachusetts Institute of Technology, Stanford, Columbia, and Southern California.[6]

Students usually would like to know exactly what it takes to get into graduate school and soon discover that nowhere are any clear, hard, fast rules written down. It is generally assumed that a "B" average is

---

[6] *Engineering Degrees—1972* (New York: Engineering Manpower Commission, November 1972).

necessary, but even this may turn out to be a flexible requirement, depending upon the school and the individual case. The ASEE's *A Report on the Education and Training of Professional Engineers in the United States*[7] offers about the clearest information available:

> To qualify for admission to study leading to a master's degree, a strong postgraduate university may require an applicant to have a standing in the upper quarter of his class, if the undergraduate school is considered relatively weak. However, a graduate school of modest standards may admit a man from a strong undergraduate school if he stood in the upper three quarters of his class. A master's degree usually requires a minimum of one academic year, although a year and a half or even two years may be required if the student must make up any deficiencies.
>
> Candidates for doctoral degrees have rarely ranked below the upper quarter of their undergraduate classes, and probably ranked in the upper five or ten percent. A minimum of three years beyond the bachelor's degree is required for a doctor's degree, although the time typically is longer—from four to six years. One of the reasons for the longer time is that doctoral candidates usually work part-time as teaching assistants or research assistants. The usual engineering doctoral degree is the Doctor of Philosophy (Ph.D.), although some schools offer the Doctor of Engineering (D. Engr.), Doctor of Engineering Science (D. Engr. Sc.), or the Doctor of Science (Sc.D.).
>
> Some schools require a thesis for the master's degree, and some do not, but a thesis is universally required for a doctor's degree. The doctoral thesis is expected to represent an original contribution to the literature, but the master's thesis more generally is expected to represent "a contribution to the training of the candidate, rather than a contribution to knowledge."[8]

In 1972, the ECPD formally began to accept applications for accreditation of advanced level programs, meaning, usually, the accreditation of master's degrees. Accreditation of master's degrees had occurred in the past, but only when the degree had been designated by a school as its "first professional degree" in that field. The difference, in 1972, was that both basic (bachelor's) and advanced (master's) levels of accreditation could be simultaneously granted in the same curriculum at a given institution. In explaining the basis for its new accreditation procedures, the ECPD stated it was only reflecting the reality of changes taking place in the engineering profession. Furthermore, said the ECPD, it was important for engineering to catch up with other professions. In

---

[7] Published by the Engineers' Council for Professional Development, New York, 1962. The passage given here has been paraphrased from this publication.

[8] The quoted passage is from *Manual of Graduate Study in Engineering.* (Washington, D.C.: ASEE, 1952), p. 19.

1900, engineering was the only profession requiring four years of collegiate study; others required less, or none. Even medicine required only three years. By 1972, virtually all of the professions, except engineering, were requiring six or more years of college preparation, such as: Master of Business Administration—six years; Master of Architecture—six years; Doctor of Jurisprudence—seven years; Doctor of Medicine—eight years.[9]

A storm of controversy arose over the new accreditation practices of the ECPD. Four of the ECPD's member societies opposed advanced level accreditation: AIAA, AIChE, AIME, and ASAE. The latter two specifically directed ECPD to continue to accredit only first professional degrees in the fields under their jurisdiction, as in the past.[10]

Somewhat surprisingly, strong opposition also arose from engineering educators. The principal reason for educators' opposition was that they feared master's-level accreditation would impose an undesirable degree of conformity on graduate programs, and they believed diversity was a quality to be sought and maintained.[11] Furthermore, they pointed out that the implicit assumption underlying any accreditation process is that the procedure is necessary to ensure quality; they declared that there was no evidence to show that quality was being inadequately maintained under past accreditation practices. In the final analysis, the dispute could only be settled by acceptance (or nonacceptance) of advanced level accreditation by individual engineering schools, and by the degree of recognition accorded to it by industry.

The Bachelor of Engineering Technology

A new kind of engineering degree has entered the picture since the late 1960s. This degree is at the baccalaureate level and usually has a name such as Bachelor of Science in Engineering Technology or, more commonly, Bachelor of Engineering Technology (B.E.T.). Interest in this kind of degree has arisen primarily because of the beliefs of some that most engineering educational programs have become very theoretical since World War II and have neglected skills formerly taught to engineers, such as drafting and manufacturing processes.

The intention of the creators of the B.E.T. degree is that its holders would occupy a middle ground between the craftsman and the engineer and would use the title "engineering technologist," rather than "engineer." In the words of the Advisory Committee for the Engineering

---

[9] *40th Annual Report, 1971–72* (New York: Engineers' Council for Professional Development, September 30, 1972), pp. 4–7.

[10] *Ibid.*, p. 20.

[11] D. C. Drucker, "Advance Level Accreditation . . . An Opposing Viewpoint," *The Registration Bulletin of the NCEE*, April 1972.

Technology Education Study: "The technologist should be a master of detail, the engineer, of the total system. . . . The development of methods or new applications is the mark of the engineer. Effective use of established methods is the mark of the technologist."[12] The principal differences in the educations of the engineer and the technologist are that the technologist is exposed to less rigor in mathematics and science than the engineer, and correspondingly gives greater attention to the development of technical skills. The ECPD accredits B.E.T. degrees, as well as B.S. degrees in engineering, but uses different criteria for the two kinds of programs.

A major question concerning B.E.T. programs is whether industrial employers will, in the limit, treat their graduates much different than bachelor-level graduates from engineering programs. Many employers have stated that they would conscientiously differentiate between engineering technologists and engineers in their job assignments, just as the framers of the B.E.T. had planned. However, by 1973, it was clear that many other employers would treat B.E.T. graduates pretty much as they would any bachelor-level engineering graduate. A comprehensive survey of B.E.T. programs in July 1972 revealed that about one half of the 3857 B.E.T. graduates in that year had received titles which classified them as some type of engineer.[13] The major distinction between B.E.T. graduates and B.S. engineering graduates in 1973 appeared to be that the former group received an average starting salary of $850 per month and the latter $930.[14]

Some confusion has surrounded the meaning of the term "engineering technology" and of a similar-sounding term, "industrial technology." The characteristics of the former have just been described; it is intended that the engineering technologist should perform in a role closely related to that of the engineer. Programs in "industrial technology," on the other hand, have been primarily outgrowths of industrial arts programs. These latter programs generally contain considerable emphasis upon business skills, and it is intended that their graduates should go into production, industrial management, and sales. But many B.S. engineering graduates go into these very same fields, and it would seem that the initially clear visions concerning the distinction of graduates of these various programs will be clouded by actual industrial practices.

A serious problem confronting holders of B.E.T. degrees is whether their educational credentials will be accepted at full value in applying for professional registration or for entrance to graduate school. The

---

[12] *Engineering Technology Education Study: Interim Report* (Washington, D.C.: American Society for Engineering Education, June 1971), p. 16.

[13] J. J. Moore and R. K. Will, "Baccalaureate Programs in Engineering Technology," *Engineering Education*, October 1973, pp. 34–38.

[14] *Job Prospects for Engineering and Technology Graduates, 1973 and 1974* (New York: Engineering Manpower Commission, October 1973).

National Council of Engineering Examiners (NCEE) at its 1972 annual meeting reported that the various state registration bodies had adopted variable practices regarding the B.E.T. Some states gave full credit for ECPD-approved B.E.T. degrees, others gave none. The Special Study Committee on Technicians and Technologists (of the NCEE) expressed its concern that variable practices by the states could endanger the foundations of comity, and that the B.E.T. degree did not contain enough theory, analysis, and design to enable a professional engineer to fulfill his public responsibility. The Special Study Committee stated that a remedy to the problem could be to encourage legislation by the states to make the holding of a B.S. degree in engineering a mandatory requirement for professional registration.[15] In somewhat similar fashion, it appeared unlikely that holders of B.E.T. degrees would be able to gain entrance to most engineering graduate programs without additional academic work because of inadequate preparation in math and science.

The professor: researcher or teacher?

A few years ago, the term "publish or perish" was rarely heard outside academic circles; today, it is practically a household term. To students, the expression implies a neglect of teaching; to parents, an unhealthy preoccupation with academic snobbery. Educators themselves are sorely split over the issue. Some are quite frank in their support of research and publication, even to the exclusion of teaching values; on the opposite side, some would like to heave research overboard and concentrate on teaching. But most teachers occupy a middle ground; they vigorously support the value of research but would like to see teaching carry more weight in the academic picture than it has in recent years.

Some students ask why research should have any position at all in universities. "Don't universities and colleges exist for the purpose of *teaching*?" they ask. The answer is that teaching is only one function that universities are expected to serve in our society. Universities serve three principal and co-equal functions: 1) the preservation of knowledge, 2) the dissemination of knowledge, and 3) the creation of new knowledge. The first function is carried out principally by the library, and the latter two—the teaching and research functions—by the faculty.

If teachers concerned themselves only with teaching what they find in textbooks, who would write the textbooks of the future? Further, it is pointed out by educators that the man who performs *only* as a lecturer invites the danger of becoming dull and repetitious. Hence, engaging in research can actually improve a professor's competence as

---

[15] *1972 Proceedings*, National Council of Engineering Examiners. (Clemson, S.C. 1973), p. 119.

a teacher. "The professor who himself is at grips with some unsolved problem as a part of his professional activity, can and does usually bring to his students a spirit of inquiry, a stimulation to tackle new ideas, and an urge to question deeper into partially understood phenomena."[16]

Of course, the hazard in all this is that research may become overemphasized. If the actions of administrative authorities lead faculty members to believe that research bears considerably more weight in advancement than does teaching, then there is a natural tendency for these professors to favor research, and teaching may suffer as a result.

The engineering professor faces an especially difficult task. First, he is usually expected to perform capably as a researcher in some branch of engineering science and to publish his results in acceptable journals. (This means he is expected to perform as a *scientist*—an *engineering* scientist.) Second, he generally must perform as a teacher of two distinct kinds of professionals, both of which are at the graduate level: he must teach 1) those who will become researchers themselves (and perhaps teachers as well) and 2) those who will do neither teaching nor research, but will become practicing professional engineers. The teaching of students of category 1) is a natural outcome of research activity and is thoroughly understood by all academic departments within a university—by physics and mathematics departments, for example. The teaching of students of category 2) is a function peculiar to professional schools, such as law, medicine, and engineering. The engineering professor must perform *all* these functions. Dean Burr of Rensselaer Polytechnic Institute has commented:

> In developing a truly professionally oriented program, a faculty faces a paradox: the fact that college professors largely further their own professional development and reputation through research. The students inevitably tend to emulate their professors. Thus the very best students in most colleges are convinced that research is the height of sophistication in engineering.[17]

Thus, the choices facing engineering faculties are not only those of research versus teaching, but also those of what *kind* of teaching. In M. P. O'Brien's words, ". . . are the schools bent on preparing students for teaching and research in universities or for the practice of engineering? . . . research in the engineering sciences must be effectively served but, in the aggregate, the engineering schools will harm this

---

[16] H. L. Hazen, "The ECPD Accreditation Program," *Journal of Engineering Education*, October 1954, p .106.

[17] A. A. Burr, "Problems of Developing a Professional-School Program in Engineering," *Journal of Engineering Education*, June 1965, p. 289.

country's ability to sustain and increase productivity if they do not develop engineer-designers of superior ability."[18]

Continuing education

Soon after an engineer has graduated, he usually begins to feel that his education was lacking in certain respects. A Purdue University study shows that the kind of lack that is sensed is fairly categorizable by age groups; thus:

> Those who have been out five years or less wish they had taken more courses of a practical nature.
>
> Those who have been out five to fifteen years wish they had taken more math and science.
>
> Those who have been out between fifteen and twenty-five years wish they had taken more courses in business and management.
>
> Those who have been out more than tweny-five years wish they had taken more humanities and fine arts.[19]

Whatever the stimulus, there is a tremendous demand today for continuing education for engineers. Part of the demand stems from the rapid changes in technology, a factor that has led to the popular but inaccurate cliché that "the half-life of an engineering education is ten years." The implication is that half of what an engineer learns in school today will be obsolete in ten years. This is nonsense, provided the man's education emphasized *fundamentals*. Fundamentals do not decay at that rate, although other things may. A man's own ability decays, unless kept alive by exercise; the demand for a certain special kind of skill may decay; the level of competence for the entire profession may move upward, and this may cause a given individual to experience relative "decay" if he doesn't keep up. In addition, a great deal of new scientific knowledge is continually being generated. As a result, the job of "keeping up" is a never-ending one, and engineers are coming more and more to accept the idea that a major portion of their time—throughout their entire careers—will be engaged in learning.

Continuing education for engineers is predominately of three types:

---

[18] M. P. O'Brien, "Professional Graduate Study in Engineering," *Journal of Engineering Education*, March 1961, pp. 579–580.

[19] G. M. Nordby, "Where Is Engineering Education Today?" *Civil Engineering*, February 1965, p. 56.

1. Specific and detailed courses on the performing of certain professional functions, for instance, computer programming
2. Courses in new technology, so that a man can cope with a declining demand for his current skills by learning some new ones. One example is the shift to integrated electronic circuits, as opposed to those made by assembling components
3. General up-grading courses, which bring a man to a higher technical level

Engineers who take courses in categories 1) and 2) generally do so because of specific needs in their jobs, and so, usually are unconcerned about receiving graduate credit. Frequently, courses of this type are given "in-house" by employers or may be given in the popular "short-course" format by universities. (In a "short-course," an overview of a subject, presented by leading authorities in the field, is packed into a course which runs full-time for a week or two weeks). Those who pursue work in the third category described above are usually anxious to have their studies lead to a master's degree. With the new emphasis on the master's degree caused by the Goals Report, this last trend can be expected to intensify. For the next decade or two, we can expect to see tens of thousands of engineers engaged in part-time study leading to advanced degrees.

The phenomenon of technical obsolescence in older engineers is especially worrisome. As long as the engineering profession was expanding rapidly, then it tended to be made up of relatively young men, and the problem of obsolescence with age could be swept under the rug. But, as the growth of the profession slowed, the age of the "average engineer" began to rise. Thus, the problem of obsolescence advanced to front rank. Research and development laboratories especially became worried. An extensive study of policies in 17 such organizations showed that there was a general fear that organizational productivity would decline as the average age of their technical staffs increased. All of them had become accustomed to perpetual growth, which automatically ensured that their staffs remained young. None of them seemed prepared to face steady state, with its implications of "corrective" layoffs in order to make way for younger people. University faculties faced identically the same problems. The only acceptable alternatives in view seemed to be 1) financial encouragement toward early retirement; 2) continuing education.[20]

There seems to be little hard evidence that productivity truly does decline with age, even though the conventional folklore asserts that it does. In the study of R & D organizations just cited, some evidence

---

[20] C. M. Van Atta, W. D. Decker, and T. Wilson, *Professional Personnel Policies and Practices of R&D Organizations* (Livermore, Calif.: Lawrence Livermore Laboratory, University of California, December 6, 1971).

emerged that scientific productivity may actually increase after age 50, although the productivity may consist more of things like pulling together the ideas of one's life work, than coming up with major new ideas. Also, the same study implied that engineers (as opposed to scientists) might actually improve with age, if the work depended on experience and judgment and not so much on creativeness. However, in another study, the investigators found definite evidence of declining performance with age, based upon evaluation of engineers by their managers. An especially significant finding in this study was that the routine taking of courses, for the purpose of continuing education, seemed to have no effect on performance. The investigators found another result which they thought was of special importance: engineers with advanced degrees were considered productive up to 10 years longer than those with bachelor's degrees. Hence, the investigators made the recommendation that mid-career graduate work intensive enough to result in a degree might effectively prolong an engineer's productive life. To accomplish such a result would require company cooperation by means of released time to attend classes, participation in "live" TV educational systems designed to bring graduate courses "in-house," and provision of "sabbatical" leaves for purposes of self-renewal.[21] Even though such programs would cost money, the engineers' employers would likely lose even more if their technical personnel were not kept productive.

The big advantage of a live TV system is that working engineers can walk a short distance to a classroom at their place of work, take the class, and be back at work immediately afterward. No time is lost in commuting to a university campus, yet the working engineer participates in a "live" classroom experience with full-time graduate students who are simultaneously sitting in a classroom on campus. The engineer at the remote location can even ask questions during class, by means of microwave links or leased telephone lines. Such systems have been very successful at many locations in the United States, but do require the active cooperation of the engineers' employers.[22]

Some new engineering fields

Biomedical engineering is a relatively new field, although it has origins extending back into the 1940s. It consists essentially of three segments: 1) the application of engineering concepts to biological phenomena—a basic research activity sometimes referred to simply

---

[21] G. W. Dalton and P. H. Thompson, "Accelerating Obsolescence of Older Engineers," *Harvard Business Review*, September–October 1971, pp. 57–67.

[22] H. H. Loomis, Jr., and H. Brandt, "Television as a Tool in Off-Campus Engineering Education," *IEEE Transactions on Education*, May 1973, pp. 101–109.

as "bioengineering"; 2) the application of engineering to the development of instrumentation, materials, artificial organs and the like, frequently called "medical engineering"; 3) the application of engineering to the improvement of health delivery systems, sometimes called "clinical engineering." Historically, medical engineering is the oldest, whereas clinical engineering is expected to come into its own during the 1970s. Preparation for these fields is likely to require graduate work. "Hybrid" programs at the Ph.D. level, having equal emphasis in engineering and in the life sciences, have been developed for the research-type biomedical engineer.[23]

Environmental engineering is new only in the sense that it is all-encompassing, including the three elements that make up our physical environment: air quality, water quality, and land quality. The water quality portion of environmental engineering is closely related to older programs in sanitary engineering. The 1972 report of ECPD listed 14 accredited environmental engineering programs, and six accredited programs in sanitary engineering—all of them at the master's level. Environment engineers may work for government agencies (federal, state, or local), universities, consulting firms, or manufacturing companies. Thus, they may work in pollution abatement and enforcement, teaching, research, testing, treatment plant design and construction, or process modification to minimize emissions.[24] They are typically concerned with air pollution, water pollution, solid waste management, radiological hazard control, pesticide hazard control, and water supply. They are also concerned with patterns of land use and with the resulting impact upon the environment.

Forest engineering as of 1973 had not yet appeared as a separate accredited engineering program in the United States, but generally was treated as an option within accredited agricultural engineering programs. Besides completion of a program in agricultural engineering, the student is expected to take a substantial amount of work in forestry. The purpose of a forest engineering program is to prepare the graduate to work in forest management, with emphasis upon physical installations and forest materials management. Included in this are such matters as tree nursery practices, reforestation, brush control, waste management, recreational development, soil and water conservation, and esthetics. The purpose throughout is to develop systems to maintain forest lands in permanent production, help to produce a desirable environment, and ensure a supply of wood products to society. Forest engineers are quick to point out that forests represent one of our few *renewable* natural resources, and thus deserve special attention.

---

[23] R. Plonsey, "New Directions for Biomedical Engineering," *Journal of Engineering Education*, December 1973, pp. 177–179.

[24] *Manpower Needs in Environmental Engineering.* Third National Conference on Environmental Engineering, July 25, 1973.

Appendix

Appendix

a. Some definitions

The following four definitions are reproduced by permission, from *Webster's Third New International Dictionary*, Copyright 1961, by G. & C. Merriam Co., Publishers of the Merriam-Webster Dictionaries.

engineering: The science by which the properties of matter and the sources of energy in nature are made useful to man in structures, machines and products.

research: Studious inquiry or examination; *esp:* critical and exhaustive investigation or experimentation having for its aim the discovery of new facts and their correct interpretation, the revision of accepted conclusions, theories or laws in the light of newly discovered facts, or the practical applications of such new or revised conclusions, theories, or laws.

science: Accumulated and accepted knowledge that has been systematized and formulated with reference to the discovery of general truths or the operation of general laws.

scientist: One learned in science and *esp.* natural science: a scientific investigator ("what distinguishes the scientist is his ability to state problems, to frame questions, so that the technicians can make the machines yield facts that are significant"—W. A. L. JOHNSON.)

The following definitions are from the Model Law, as prepared by the National Council of Engineering Examiners. The Model Law serves merely as a guide to law-making bodies and has no legal effect unless written into law by a legislative body.

engineer: The term, "engineer," within the intent of this Act shall mean a person who, by reason of his special knowledge and use of the mathematical, physical, and engineering sciences and the principles and methods of engineering analysis and design, acquired by engineering education and experience, is qualified to practice engineering.

practice of engineering: The term, "practice of engineering," within the intent of this Act, shall mean any service or creative work, the adequate performance of which requires engineering education, training, and experience in the application of special knowledge of the mathematical, physical, and engineering sciences to such services or creative work as consultation, investigation, evaluation, planning and design of engineering works and systems, planning the use of land and water, teaching of advanced engineering subjects, engineering surveys, and the inspection of construction for the purpose of assuring compliance with drawings and specifications; any of which embraces such service or work either public or private, in connection with any utilities, structures, buildings, machines, equipment, processes, work systems, projects, and industrial or consumer products or equipment of a mechanical, electrical, hydraulic, pneumatic or thermal nature, insofar as they involve safeguarding life, health or property, and including such other professional services as may be necessary to the planning, progress and completion of any engineering services.

The following definition was prepared by the Engineers' Council for Professional Development (ECPD).[1]

engineering: Engineering is the profession in which a knowledge of the mathematical and natural sciences gained by study, experience and practice is applied with judgment to develop ways to utilize, economically, the materials and forces of nature for the benefit of mankind.

b. Profession and professional practitioners

The following material also was prepared by the ECPD.[2]

Attributes of a profession and its practitioners

Of a profession:

1. It must satisfy an indispensable and beneficial social need.
2. Its work must require the exercise of discretion and judgment and not be subject to standardization.
3. It is a type of activity conducted upon a high intellectual plane. (a) Its knowledge and skills are not common possessions of the general public; they are the results of tested research and experience and are acquired through a special discipline of education and practice. (b) Engineering requires a body of distinct knowledge (science) and art (skill).
4. It must have group consciousness for the promotion of technical knowledge and professional ideals and for rendering social services.

---

[1] *Annual Report, 1963* (New York: ECPD, 1963), p. 3.
[2] *Annual Report, 1945* (New York: ECPD, 1945).

5. It should have legal status and must require well-formulated standards of admission.

Professional practitioners:

1. They must have a service motive, sharing their advances in knowledge, guarding their professional integrity and ideals, and tendering gratuitous public service in addition to that engaged by clients.
2. They must recognize their obligations to society and to other practitioners by living up to established and accepted codes of conduct.
3. They must assume relations of confidence and accept individual responsibility.
4. They should be members of professional groups and they should carry their part of the responsibility of advancing professional knowledge, ideals, and practice.

c. Canons of ethics

After several years of review, the ECPD in 1973 published the following revised set of Canons of Ethics[3] and urged that they be adopted by its constituent societies:

The fundamental principles

The Engineer upholds and advances the honor and dignity of the engineering profession by:

1. Being honest and impartial, and serving with fidelity his employer, his clients, and the public
2. Striving to increase the competence and prestige of the engineering profession
3. Using his knowledge and skill for the enhancement of human welfare
4. Supporting the professional and technical societies of his discipline.

The fundamental canons

1. The Engineer has proper regard for the safety, health, and welfare of the public in the performance of his professional duties.
2. The Engineer does everything in his power to assure the safety and reliability of products for which he is responsible.

---

[3] *41st Annual Report, 1972–73* (New York: ECPD, 1973), pp. 24–28.

3. The Engineer is guided in all his professional relations by the highest standards of integrity, and acts in professional matters for each employer or client as a faithful agent or trustee.
4. The Engineer endeavors to extend public knowledge, and to prevent misunderstanding of the achievements of engineering.
5. The Engineer undertakes engineering assignments for which he will be responsible only when qualified by training and experience.
6. The Engineer does not associate with nor allow the use of his name by an enterprise of questionable character, nor does he become professionally associated with those who do not conform to ethical practices, nor with persons not professionally qualified to render the services for which the association is intended.
7. The Engineer upholds the principle of appropriate and adequate compensation for those engaged in engineering work.
8. The Engineer accepts compensation, financial or otherwise, from only one interested party for the same service, or for services pertaining to the same work, unless there is full disclosure to and consent of all interested parties.
9. The Engineer cooperates in extending the effectiveness of the profession by interchanging information and experience with engineers and students, and he endeavors to provide opportunities for the professional development and advancement of those under his supervision.
10. The Engineer does not disclose confidential information concerning the business affairs or technical processes of any present or former employer or client or bidder under evaluation without his consent.
11. The Engineer gives credit for engineering work to those to whom credit is due, and recognizes the proprietary interests of others.
12. The Engineer does not unfairly compete with another engineer.
13. The Engineer does not maliciously or falsely, directly or indirectly, injure the professional reputation, prospects, practice nor employment of another engineer, nor does he indiscriminately criticize another engineer's work.
14. The Engineer advertises his work or merit only in a dignified manner, and avoids conduct or practice likely to discredit or unfavorably reflect upon the dignity or honor of the profession.

Guidelines for use with the fundamental canons of ethics

1.a. The Engineer shall regard the public welfare as of primary importance.

b. He shall not complete, sign, or seal plans and/or specifications that are not of a design safe to the public health and welfare and in conformity with accepted engineering standards. If the employer or client insists on such unprofessional conduct, he shall notify those with power to take corrective action. If proper action is not taken, he shall withdraw from further service on the project.

c. He should seek opportunities to be of constructive service in civic affairs and work for the advancement of the safety, health and well-being of his community.

2.a. The Engineer does whatever he can to provide published standards, test codes and quality control procedures that will enable the public to understand the degree of safety or life expectancy associated with the use of the products for which he is responsible.

b. The Engineer will conduct a review of the safety and reliability of the product or system for which he is responsible before giving his approval to the plans for the design.

c. The Engineer will advise his employer if an adequate review of the safety and reliability of the product or system has not been made or if the design imposes hazards to the public through its use.

d. The Engineer will withhold his approval of a product or system if changes or modifications are made which would affect adversely its performance insofar as safety and reliability are concerned.

3.a. The Engineer shall be realistic and honest in all estimates, reports, statements, and testimony.

b. He shall admit and accept his own errors when proven wrong and refrain from distorting or altering the facts in an attempt to justify his decision.

c. He shall advise his employer or client when he believes a project will not be successful.

d. He shall not accept professional employment outside of his regular work or interest without the consent of employer.

e. The Engineer shall inform his employer or client of any business connections, interests, or other circumstances which may be deemed as influencing his judgment or the quality of his services to his employer or client.

f. When in public service as a member, advisor or employee of a governmental body or department, an engineer shall not participate in considerations or actions with respect to services provided by him or his organization in private engineering practice.

g. An Engineer shall not solicit or accept an engineering contract from a governmental body on which a principal or officer of his organization serves as a member unless such principal or officer disqualifies himself for action on that contract.

h. He shall not attempt to attract an engineer from another employer by unfair methods.

i. He shall engage, or advise engaging, experts and specialists, when in his judgment such services are in his employer's or client's best interests.

4.a. The Engineer will express an opinion on an engineering subject only when founded on adequate knowledge and honest conviction.

b. He shall insist on the use of facts in reference to an engineering project in a group discussion, public forum or publication of articles.

c. He shall preface any partisan statements, criticisms, or arguments that he may issue by clearly indicating on whose behalf they are made.

d. He shall be dignified and modest in explaining his work and merit, and shall ever uphold the honor and dignity of his profession.

5.a. If his engineering judgment is overruled by nontechnical authority, he will clearly point out the probable consequences.

6.a. The Engineer shall conform with all applicable laws and governmental regulations in his practice of engineering.

b. He shall not use association with a nonengineer, a corporation, or partnership, as a "cloak" for unethical acts, but shall accept personal responsibility for his professional acts.

c. He will not sign or seal plans and/or specifications not prepared by him or under his direct supervision.

7.a. The Engineer shall not undertake or agree to perform any engineering service on a free basis, except professional services which are advisory in nature for civic, charitable, religious, or eleemosynary non-profit organizations.

b. He shall not undertake work at a fee or salary below the accepted practice of the profession in the area.

c. When hiring other engineers, he shall offer a salary according to the engineer's qualifications and the recognized standards in the particular geographical area.

d. If in sales employ, he shall not offer, nor give engineering consultation, nor designs other than specifically applying to the equipment being sold.

8.a. The Engineer shall not accept financial nor other considerations, including free engineering designs, from material or equipment suppliers for specifying their product.

b. He shall not accept commissions nor allowances, directly or indirectly, from contractors or other parties dealing with his employer or clients in connection with work for which he is responsible.

9.a. The Engineer should encourage his engineering employees to further their education.

b. He should encourage engineering employees to attend and present papers at professional and technical society meetings.

c. He should encourage his qualified engineering employees to become registered at the earliest possible date.

d. He shall assign a professional engineer duties of a nature to utilize his full training and experience, insofar as possible, and delegate lesser functions to subprofessionals or to technicians.

e. He shall provide a prospective engineering employee with complete information on working conditions and his proposed status of employment, and after employment will keep him informed of any changes in them.

10.a. The Engineer, while in the employ of others, shall not enter promotional efforts nor negotiations for work or make arrangements for other employment as a principal or to practice in connection with a specific project for which he has gained particular and specialized knowledge without the consent of all interested parties.

b. The Engineer shall treat information coming to him in the course of his assignment as confidential and shall not use such information as a means of making personal profit if such action is adverse to the interest of his employer, his client, or the public.

c. The Engineer shall treat as confidential findings of an engineering commission or board of which he is a member except with official consent.

11.a. Whenever possible, he shall name the person or persons who may be individually responsible for designs, inventions, writings, or other accomplishments.

b. When an Engineer uses designs supplied to him by a client, the designs remain the property of the client and shall not be duplicated by the Engineer for others without express permission.

c. Before undertaking work for others in connection with

which he may make improvements, plans, designs, inventions, or other records which may justify copyrights or patents, the Engineer shall enter into a positive agreement regarding the ownership.

d. Designs, data, records, and notes made by an Engineer and referring exclusively to his employer's or client's work are his employer's or client's property.

12.a. The Engineer shall not attempt to supplant another engineer in a particular employment after becoming aware that definite steps have been taken toward the other's employment or after he has been employed.

b. He shall not pay nor offer to pay, either directly or indirectly, any commission, political contribution or make a gift or other consideration in order to secure work, exclusive of securing salaried positions through employment agencies.

c. The Engineer should negotiate a method and rate of compensation commensurate with the agreed upon scope of services. A meeting of the minds of the parties to the contract is essential to mutual confidence. The public interest requires that the cost of engineering services should not be a controlling consideration.

These principles shall be applied by the Engineer in obtaining the services of other professionals.

When engaged in work in foreign countries, the Engineer shall make every reasonable effort to seek a change in the procedure in accordance with this section, but if this is not successful, the Engineer may conform to the procedures as required by the laws, regulations or practices of the foreign country.

d. An Engineer shall not request, propose nor accept a professional commission on a contingent basis under circumstances under which his professional judgment may be compromised, or when a contingency provision is used as a device for promoting or securing a professional commission.

e. An Engineer shall not use equipment, supplies, laboratory nor office facilities of his employer to carry on outside private practices without consent.

13.a. An Engineer in private practice shall not review the work of another engineer for the same client, except with the knowledge of such engineer, or unless the contractual agreement for the work has been terminated.

b. An Engineer in governmental, industrial or educational employ is entitled to review and evaluate the work of other engineers when so required by his employment duties.

c. An Engineer in sales or industrial employ is entitled to make engineering comparisons of his products with products of other suppliers.

d. An Engineer who has proof that an Engineer has been unethical, illegal or unfair in his practices should advise proper professional authority within the engineering profession.

14.a. An Engineer shall not advertise his professional services but may utilize the following means of identification:

(1) Professional cards and listings in recognized and dignified publications, provided they are consistent in size and are in a section of the publication regularly devoted to such professional cards and listings. The information displayed must be restricted to firm name, address, telephone number, appropriate symbol, names of principal participants and the fields of practice in which the firm is qualified.

(2) Signs on equipment, offices and at the site of projects for which he renders service, limited to firm name, address, telephone number and type of service, as appropriate.

(3) Brochures, business cards, letterheads and other factual representations of experience, facilities, personnel and capacity to render service, providing the same are not misleading relative to the extent of participation in the projects cited.

(4) Listings in the classified section of telephone directories, limited to name, address, telephone number and specialties in which the firm is qualified.

b. An Engineer may use display advertising in recognized dignified business and professional publications, providing it is factual, free from ostentation, contains no laudatory expressions nor implication and is not misleading with respect to the Engineer's extent of participation in projects described.

c. An Engineer may prepare articles for the lay or technical press which are factual, dignified and free from ostentations or laudatory implications. Such articles shall not imply other than his direct participation in the work described unless credit is given to others for their share of the work.

d. An Engineer may extend permission for his name to be used in commercial advertisements, such as may be published by manufacturers, contractors, material suppliers, etc., only by means of a modest dignified notation acknowledging his participation and the scope thereof in the project or product described. Such permission shall not include public endorsement of proprietary products.

e. An Engineer will not allow himself to be listed for employment using exaggerated statements of his qualifications.

f. An Engineer may advertise for recruitment of personnel in appropriate publications or by special distribution. The information presented must be displayed in a dignified manner, restricted to firm name, address, telephone number, appropriate symbol, names of principal participants, the fields of practice in which the firm is qualified and factual descriptions of positions available, qualifications required and benefits available.

d. Faith of the engineer[4]

I AM AN ENGINEER. In my profession I take deep pride, but without vain-glory; to it I owe solemn obligations that I am eager to fulfill.

As an Engineer, I will participate in none but honest enterprise. To him that has engaged my services, as employer or client, I will give the utmost of performance and fidelity.

When needed, my skill and knowledge shall be given without reservation for the public good. From special capacity springs the obligation to use it well in the service of humanity; and I accept the challenge that this implies.

Jealous of the high repute of my calling, I will strive to protect the interests and the good name of any engineer that I know to be deserving; but I will not shrink, should duty dictate, from disclosing the truth regarding anyone that, by unscrupulous act, has shown himself unworthy of the profession.

Since the Age of Stone, human progress has been conditioned by the genius of my professional forebears. By them have been rendered usable to mankind Nature's vast resources of material and energy. By them have been vitalized and turned to practical account the principles of science and the revelations of technology. Except for this heritage of accumulated experience, my efforts would be feeble. I dedicate myself to the dissemination of engineering knowledge, and, especially to the instruction of younger members of my profession in all its arts and traditions.

To my fellows I pledge, in the same full measure I ask of them, integrity and fair dealing, tolerance and respect, and devotion to the

---

[4] Prepared by the Ethics Committee, Engineers' Council for Professional Development.

standards and the dignity of our profession; with the consciousness, always, that our special expertness carries with it the obligation to serve humanity with complete sincerity.

e. Guidelines to professional employment for engineers and scientists[5]

These guidelines have been endorsed by the following engineering societies:

American Association of Cost Engineers
American Institute of Chemical Engineers
American Institute of Industrial Engineers
American Nuclear Society
American Society of Civil Engineers
American Society of Mechanical Engineers
American Society for Quality Control
Engineers' Council for Professional Development
Engineers Joint Council
Institute of Electrical and Electronics Engineers
Institute of Traffic Engineers
Instrument Society of America
National Institute of Ceramic Engineers
National Society of Professional Engineers
Society of Fire Protection Engineers

Foreword

This publication is a guide to mutually satisfying relationships between professional employees and their employers. In this document, professional employees are defined as engineers and scientists. These guidelines cover factors peculiar to professional employment, and omit many generally accepted precepts of personnel relations which are common to all classifications of employees.

These guidelines are applicable to professional employment in all fields and in all areas of practice (including both nonsupervisory and supervisory positions), and are based on the combined experience and judgment of all of the endorsing societies.

It must be stressed in the implementation of these guidelines that they represent desirable general goals rather than a set of specific minimum standards. Wide variations in circumstances and individual organizational practices make it inappropriate to judge any given employer on the basis of any single employment policy or fringe benefit.

---

[5] First edition, January 1, 1973.

Rather, attention should be devoted to evaluating the entire employment "package," including such intangibles as opportunity for future advancement or participation in profits, location, local cost of living, and other factors which may be important to professional employees.

Observance of the spirit of these guidelines will minimize personnel problems, reduce misunderstandings, and generate greater mutual respect. It is anticipated that they will be of use to employers in evaluating their own practices, to professional employees in evaluating both their own responsibilities and those of their employers, and to new graduates and other employment seekers in obtaining a better picture of prospective employers. Where differences in interpretation occur, they may be referred to the headquarters office of any of the endorsing societies.

Objectives

The endorsing societies, with their avowed purpose to serve the public and their professions, recognize clearly that in order to make a maximum contribution, it is necessary for professional employees and employers to establish a climate conducive to the proper discharge of mutual responsibilities and obligations.

Essential and prerequisite to establishing such a climate are:

1. Mutual loyalty, cooperation, fair treatment, ethical practices, and respect are the basis for a sound relationship between the professional and his employer.
2. The professional employee must be loyal to the employer's objectives and contribute his creativity to those goals.
3. The responsibility of the professional employee to safeguard the public interest must be recognized and shared by the professional employee and employer alike.
4. The professional growth of the employee is his prime responsibility, but the employer undertakes to provide the proper climate to foster that growth.
5. Factors of age, race, religion, political affiliation, or sex should not enter into the employee/employer relationship.

Effective use of these guidelines is accomplished when the employer provides each present and prospective professional employee with a written statement of his policies and practices relating to each of the items covered. Adherence to these guidelines by employers and professional employees will provide an environment of mutual trust and confidence. Local conditions may result in honest differences in interpretation of, and in deviation from, the details of these guidelines. Such differences should be resolved by discussions leading to an understanding which meets the spirit of the guidelines.

I. Recruitment

Employment should be based solely on professional competence and ability to adequately perform assigned responsibilities, with employee qualifications and employment opportunities represented in a factual and forthright manner. The employer's offer of employment and the employee's acceptance should be in writing, including a clear understanding with regard to relocation assistance; past, present, and future confidentiality and patent obligations; salary; expected duration of employment; and other relevant employment conditions and benefits.

Professional Employee

1. The professional employee (applicant) should attend interviews and accept reimbursement only for those job opportunities in which he has a sincere interest. The applicant should prorate costs for multiple interviews during a given trip on a rational basis. The guiding principles should be that the applicant receives neither more nor less than the cost of the total trip.
2. The applicant should carefully evaluate past, present, and future confidentiality obligations in regard to trade secrets and proprietary information connected with the potential employment. He should not seek or accept employment on the basis of using or divulging any trade secrets or proprietary information.
3. Having accepted an offer of employment, the applicant is morally obligated to honor his commitment unless formally released after giving adequate notice of intent.
4. The applicant should not use the funds or time of his current employer for the purpose of seeking new employment unless approved by the current employer.

Employer

1. The policy of the employer regarding payment of expenses incurred by the applicant in attending the interview must be made clear prior to the arranged interview.
2. The applicant should have an interview with his prospective supervisor in order to understand clearly the technical and business nature of the job opportunity. This prospective supervisor should be ethically responsible for all representations regarding the conditions of employment.
3. Applications for positions should be confidential. The expressed consent of the applicant should be obtained prior to communicating with a current employer.
4. Employers should minimize hiring during periods of major cur-

tailment of personnel. Hiring of professional personnel should be planned at all times to provide satisfying careers.

5. Agreements among employers or between employer and professional employee which limit the opportunity of professional employees to seek other employment or establish independent enterprises are contrary to the spirit of these guidelines.
6. Having accepted an applicant, an employer who finds it necessary to rescind an offer of employment should make adequate reparation for any injury suffered.

II. Terms of employment

Terms of employment should be in writing, in accordance with the applicable laws, and consistent with generally accepted ethical professional practices.

Professional Employee

1. The professional employee should be loyal to his employer. He should accept only those assignments for which he is qualified; should diligently, competently, and honestly complete his assignments; and he should contribute creative, resourceful ideas to his employer while making a positive contribution toward establishing a stimulating work atmosphere and maintaining a safe working environment.
2. The professional employee should have due regard for the safety, life, and health of the public and fellow employees in all work for which he is responsible. Where the technical adequacy of a process or product is involved, he should protect the public and his employer by withholding approval of plans that do not meet accepted professional standards and by presenting clearly the consequences to be expected if his professional judgment is not followed.
3. The professional employee should be responsible for the full and proper utilization of his time in the interest of his employer and the proper care of the employer's facilities.
4. The professional employee should avoid any conflict of interest with his employer, and should immediately disclose any real or potential problem which may develop in this area. He should not engage in any other professional employment without his employer's permission.
5. The professional employee should not divulge technical proprietary information while he is employed. Furthermore, he should not divulge or use this information for an agreed upon period after employment is terminated.
6. The professional employee should only sign or seal plans or

specifications prepared by himself or others under his supervision, or plans or specifications that he has reviewed and checked to his personal satisfaction.

7. The professional employee should not accept payments or gifts of any significant value, directly or indirectly, from parties dealing with his client or employer.

Employer

1. The employer should inform his professional employees of the organization's objectives, policies, and programs on a continuing basis.
2. The professional employee should receive a salary in keeping with his professional contribution which reflects his abilities, professional status, responsibility, the value of his education and experience and the potential value of the work he will be expected to perform. The salary should be commensurate with the salaries of other employees both professional and nonprofessional. Sound indirect compensation programs should be provided. The most important are retirement plans, health and life insurance, sick leave, paid holidays and paid vacations.
3. The employer should establish a salary policy, taking into account published salary surveys, and provide equitable compensation for each employee commensurate with his position and performance. The salary structure should be reviewed annually to keep the assigned dollar values adjusted to the current economy.
4. Each individual position should be properly classified as to its level in the overall salary structure. The evaluation of each position should consider such factors as the skill required for acceptable performance, the original thinking required for solving the problems involved, and the accountability for an action and its consequences.
5. Economic advancement should be based upon a carefully designed performance review plan. Provisions should be made for accelerated promotions and extra compensation for special accomplishments. At least annually, performance evaluations and salary reviews should be conducted for the individual professional employee by his supervisor. Performance evaluations should include discussion on how well he has performed his work and what he can do to improve. The professional employee should be clearly informed if his performance is considered unsatisfactory. All promotions in salary and responsibility should be on an individual merit basis.
6. For the professional employee whose aptitude and interests are technical rather than supervisory, equivalent means of advancement and recognition should be provided.

7. It is inappropriate for a professional employee to use a time clock to record arrival and departure, particularly since situations may arise which require unusual effort on his part. However, if the work demanded of a professional employee regularly exceeds the normal working hours for extended periods, the employer should compensate him for his continuing extra effort according to a clearly stated policy.
8. The professional employee should be included in an adequate pension plan which provides for early vesting of rights in safeguarded pension funds. Vesting should be so scheduled that it does not seriously affect either the employer's or the professional employee's decision as to continued employment. As a goal, eligibility for participation should not exceed one year after employment, maximum full vesting time should be five years, and the minimum pension upon reaching retirement should be no less than 50 percent of the average best five years' salary (based on a 40-year working career with a single employer). If a pension plan is not provided or the benefits are less than outlined above, other compensation should be increased proportionately.
9. The employer should provide office, support staff and physical facilities which promote the maximum personal efficiency of the professional employee.
10. Duties, levels of responsibility, and the relationship of positions within the organizational hierarchy should be clearly defined and should be accurately reflected in position titles.
11. The employer should not require the professional employee to accept responsibility for work not done under his supervision.
12. The employer should provide formal assurance through organizational policy that it will defend any suits or claims against individual professional employees of the organization in connection with their authorized professional activities on behalf of the employer.
13. There should be no employer policy which requires a professional employee to join a labor organization as a condition of continued employment.
14. It is the employer's responsibility to clearly identify proprietary information.

III. Professional development

The employee and the employer share responsibility for professional development of the employee—the employee to establish the goals and take the initiative to reach them, and the employer to provide the environment and attitude which is conducive to profesional growth.

Professional Employee

1. Each professional employee is responsible for maintaining his technical competence and developing himself through a program of continuing education.
2. The professional employee should belong to and participate in the activities of appropriate professional societies in order to expand his knowledge and experience. Such participation should include the preparation of professional and technical papers for publication and presentation.
3. The professional employee should achieve appropriate registration and/or certification as soon as he is eligible.
4. The professional employee should recognize his responsibility to serve the public by participating in civic and political activities of a technical and nontechnical nature. Such participation, however, should be undertaken solely as a responsibility of the individual without interfering with the timely execution of his work and without involving the employer.

Employer

1. The employer, as a matter of policy, should provide an atmosphere which promotes professional development. This will include, among other programs, encouraging and supporting membership and attendance at professional society meetings and at formal courses of study which will enable the employee to maintain his technical competence.
2. The employer should consider compensated leaves of absence for professional study as a means of enabling the employee to improve his competence and knowledge in a technical field.
3. Consistent with employer objectives, the employee should be given every opportunity to publish his work promptly in the technical literature and to present his findings at technical society meetings.
4. It is in the best interest of the employer to encourage continuing education to broaden the qualifications of employees through self-improvement, in-house programs, formal education systems in the institutions of higher learning, and meetings and seminars on appropriate subjects.
5. The employer should encourage and assist professional employees to achieve registration and/or certification in their respective fields.

IV. Termination and transfer

Adequate notice of termination of employment should be given by the employee or employer as appropriate.

Professional Employee

1. If the professional employee decides to terminate his employment, he should assist the employer in maintaining a continuity of function, and he should provide at least one month's notice. When termination is initiated by the employee, no severance pay is due.

Employer

1. Additional notice of termination, or compensation in lieu thereof, should be provided by the employer in consideration of responsibilities and length of service. As a desirable goal, permanent employees (after initial trial period) should receive notice or equivalent compensation equal to one month, plus one week per year of service. In the event that the employer elects notice in place of severance compensation, then he should allow the employee reasonable time and facilities to seek new employment.
2. Employers should make every effort to relocate terminated professional employees either within their own organizations or elsewhere. Consideration should be given to continuing major employee protection plans for some period following termination, and to their full reinstatement in the event of subsequent reemployment.
3. If a professional employee is involuntarily terminated on the basis of early retirement, the employer should consider an equitable provision for an adequate income for the period remaining until the employee receives his pension at his normal retirement age.
4. In a personal interview, the employer should inform the employee of the specific reasons for his termination.
5. The employer should provide an adequate transfer-time notice, with due consideration to the extent of personal matters which the professional employee must settle before moving. All normal costs of the transfer should be paid by the employer including moving expenses, realtor fees, travel expenses to the new location to search for housing, and reasonable living expenses for the family until permanent housing is found. Unusual moving expense reimbursement should be settled in a discussion between the employee and employer.

This document is subject to periodic review by the participating societies for the purpose of keeping it current. Suggested amendments will be considered collaboratively in connection with future revised editions.

Index

Index